Recipes and Everyday Knowledge

Recipes and Everyday Knowledge

Medicine, Science, and the Household in Early Modern England

ELAINE LEONG

THE UNIVERSITY OF CHICAGO PRESS   CHICAGO AND LONDON

The University of Chicago Press, Chicago 60637
The University of Chicago Press, Ltd., London
Published 2018
Printed in the United States of America

27 26 25 24 23 22 21 20 19 18   1 2 3 4 5

ISBN-13: 978-0-226-58349-5 (cloth)
ISBN-13: 978-0-226-58366-2 (paper)
ISBN-13: 978-0-226-58352-5 (e-book)
DOI: https://doi.org/10.7208/chicago/9780226583525.001.0001

Library of Congress Cataloging-in-Publication Data

Names: Leong, Elaine Yuen Tien, 1975– author.
Title: Recipes and everyday knowledge: medicine, science, and the household in early
    modern England / Elaine Leong.
Description: Chicago; London: The University of Chicago Press, 2018. | Includes
    bibliographical references and index.
Identifiers: LCCN 2018009198 | ISBN 9780226583495 (cloth: alk. paper) |
    ISBN 9780226583662 (pbk: alk. paper) | ISBN 9780226583525 (e-book)
Subjects: LCSH: Medicine, Popular—History—17th century. | Formulas, recipes, etc.—
    England—History—17th century. | Home economics—England—History—17th century. |
    Note-taking—History—17th century.
Classification: LCC R487 .L46 2018 | DDC 615.8/8—dc23
LC record available at https://lccn.loc.gov/2018009198

♾ This paper meets the requirements of ANSI/NISO Z39.48–1992 (Permanence of Paper).

FOR ALEX AND NICHOLAS

Contents

Abbreviations

BL   British Library, London

Bodleian   Bodleian Library, Oxford

CUL   Cambridge University Library, Cambridge

Folger   Folger Shakespeare Library, Washington, DC

Glasgow   Glasgow University Library, Glasgow

Huntington   Henry E. Huntington Library, San Marino, California

NLM   National Library of Medicine, Bethesda, Maryland

NYAM   New York Academy of Medicine Library, New York City, New York

NYPL   New York Public Library, New York City, New York

OED   *Oxford English Dictionary Online*, ed. J. Simpson and E. Weiner (Oxford, 2003)

Oxford DNB   *Oxford Dictionary of National Biography Online*, ed. H. C. G. Matthew and B. Harrison (Oxford, 2004)

RCP   Royal College of Physicians Library, London

RCS   Royal College of Surgeons Library, London

STC   A. W. Pollard and G. R. Redgrave, *A Short-Title Catalogue of Books Printed in England, Scotland, and Ireland and of English Books Printed Abroad, 1475–1640*, 2nd ed., rev. by K. F. Pantzer, 3 vols. (London, 1976–91)

Wellcome   Wellcome Library for the History and Understanding of Medicine, London

Wing   D. Wing, *Short-Title Catalogue of Books Printed in England, Scotland, Ireland, Wales and British America and of English Books Printed in Other Countries, 1641–1700*, 2nd ed., 3 vols. (New York: Index Committee of the Modern Language Association of America, 1972–94)

Conventions

All dates are given in the old style, except that the year is taken as beginning on 1 January. All quotations are in their original spelling with i/j and
u/v modernized and superscripts lowered. Common contractions such as
"ye," "yt," "wch," and "wt" have been expanded to "the," "it," "which,"
and "with." The place of publication for all works printed before 1900 is
London unless otherwise stated.

Recipes, Households, and Everyday Knowledge

In 1658 Sir Edward Dering (1625–1684), gentleman, politician, and poet, spent his summer making and testing medicines in his home in Pluckley, Kent. Sometime in May of that year, he started a new section in his commonplace book headed "physicall practises" and began writing down a series of recipes for medical remedies and records of his own trials of these medicines. The first entry, a remedy for the gout, recommended a simple mixture of three herbs (danewort, plantain, and nettle) applied warm to the afflicted place and having the patient fast. Dering carefully records, "This I tryed upon James Homewoods hand, my bayliffe; & he found good of it."[1] Among Dering's early adventures in testing recipes was a trial to make the spirit of honey. He wrote:

> I tooke a quart of very strong stale mead, & distilled it in Alembico thinking so to have a spirit of honey, but the water that came out was small, & running still smaller we tooke it of, & found the honey remaining in the bottome of the still stronger then it was before: contrary to other distillations: but I suppose what came first was the water which was mingled with the honey when it was made into mead at first.[2]

As the entry makes clear, things did not go exactly as planned. Instead of producing spirit of honey, Dering ended up with water and lots of "strong" honey at the bottom of his still. Comparing this particular trial with previous ones, Dering speculated that this time he had observed the separation of the mead back to its components. Although this particular trial failed to produce the intended product, Dering still saw fit to record his experiences. For him this was clearly an occasion to gain further knowledge about how substances like mead, water, and honey reacted to heat and distillation. The trial was driven both by a desire to produce the spirit of honey and by curiosity about materials, production techniques, and the natural world. Perhaps encouraged by the insight gained through this experiment, over the next few months Dering continued to try his hand at making and testing various drugs, including an elder ointment, an ointment for scalds or burns, a medicine for the gout, an oil or spirit of roses, syrups of cowslips, baume, coltsfoot, fumitory, damask roses, poppies, and maidenhair, a water for stomach ailments, angelica water, and much more.[3] It is clear that in 1658 Dering, and most probably his family and staff, had a busy summer.

Dering's enthusiasm for recipes and recipe trials was not unusual. In fact, early modern English gentlemen and gentlewomen were gripped by recipe fever. They eagerly exchanged know-how in letters, around the dinner table, and at public taverns and diligently wrote down the treasured knowledge in notebooks of all shapes and sizes. Hundreds of these handwritten books, brimming with recipes, survive in modern libraries and archives and are joined by a large number of early modern printed works, offering a plethora of know-how.[4] The central place of recipes within the cultural milieu of early modern England is demonstrated by multiple references in contemporary literature and plays. We might recall, for example, the key role Helena's fistula remedy plays in William Shakespeare's *All's Well That Ends Well* or the book of secrets and the virginity test in Thomas Middleton's *The Changeling*.[5]

The rich archive of surviving texts and the continual appearance of recipes in personal writings and in literary works attest to the importance of recipe collection and exchange as a social and cultural phenomenon in early modern England. Masters and mistresses of large households were expected to have basic knowledge of physic, cookery, and sugar craft, as outlined in many contemporary housewifery and husbandry manuals.[6] The seeking out and recording of recipe knowledge was firmly within the established practice of household management and planning. For example, in December 1647 Dering's contemporary Sir Cheney Culpeper (1601–1663) sent a panicky letter to the Anglo-Prussian intelligencer Samuel Hartlib (1600–1662). Cheney was looking for two missing recipe books and implored Hartlib to "searche diligently" for them at both his own and Mr. Woosley's lodgings, for the loss of the books seems to have caused a crisis. Culpeper's wife, Elizabeth, was ready to accuse a certain

"untowarde wenche" of having stolen them as "shee hathe done severall other bookes" and was also blaming Sir Cheney himself for "carrying
them to London." As a last-ditch attempt, Culpeper writes, "my laste
hopes are that you have copied them out, I pray ease my hearte, for my
Wife muche wantes them."[7] The desperate tone of Culpeper's letter illu-
minates not only the dynamics of the Culpeper marriage and household
but also the importance placed on recipe books and recipe knowledge by
men as well as women. Sharing recipes strengthened social and family alliances, whether as part of formal gift exchanges or in instances of mere
"borrowing" as in the case of Hartlib and Cheney. Social knowledge and
networks crucially paved the way for recipe exchange.

The cases of the Derings and the Cheneys present the early modern
household as a key setting for many recipe-related practices. This was particularly true for medicine. As part of the general overseeing and management of the household economy, gentlemen and gentlewomen dedicated
considerable time, manpower, and resources to all kinds of home-based
health care. Not only was the household considered the first resource for
dealing with many medical ailments, recent studies have demonstrated
that the domestic space was one of the main sites for medical intervention
and the promotion of health.[8] Householders were quick to combine self-
diagnosis and self-treatment with commercially available medical care,
and many produced their own homemade medicines.[9] Gathering, trying,
and testing medicines and, relatedly, foods were part of this set of activities to gather and construct knowledge about health and the body.

As shown by Edward Dering's trial with the spirit of honey, working
through these recipes let early modern men and women observe material changes, test out equipment, techniques, and production methods, and
gain further knowledge about materials and technical processes. Consequently, household recipe books record not only how householders produced a range of foods and medicines, but also how they investigated and
used natural materials and production techniques, how they understood
and looked after their bodies in sickness and health, and how they po-
sitioned themselves within their natural environment. A study of these
practices offers historians a unique entry into home-based investigations
into these areas of knowledge.

Through extensive archival work, this book elucidates the multiple,
connected ways early modern men and women sought to understand
their bodies and the surrounding natural world within their own homes.
By putting the production and circulation of recipes at the heart of what

we might call "household science"—that is, quotidian home-based investigations of the natural world—I situate these knowledge processes
within larger and current conversations in gender and cultural history, the
history of the book and archives, and the history of science, medicine, and
technology. Yet, as this book shows, household recipe practices generated
not just natural knowledge but also knowledge about materials and techniques, household management, social and family strategies, and health
and the human body. The broad term "everyday knowledge" aptly serves
as an umbrella for these varied epistemic activities, reminding us that
"knowing" often comes from the "practices of everyday life."[10]

Recipes and the Knowledge of Making

The Derings' and the Cheneys' recipe-related experiences were part of
a much larger pan-European scheme of everyday tasks directed toward
making things—whether medicines, imitation coral, colored glass, or the
elusive philosopher's stone—and recording *how* to make things in the
form of recipes. Early modern Europe was awash with recipes. Medical
practitioners of all kinds wrote down formulas and instructions for making medicaments.[11] Housewives such as Hannah Woolley (1622–c. 1675)
and chefs such as Bartolomeo Scappi (c. 1500–1577) alike noted their preferred ways to bake a cheesecake or preserve apricots.[12] London merchants such as Clement Draper (1542–1620) and Hugh Plat (1552–1608)
eagerly jotted down recipes for the philosopher's stone alongside those
for diet drinks and plague cakes.[13] Alchemists such as George Starkey
(1628–1665) had "laboratory" notebooks in which they detailed their experiments in manipulating matter.[14] Artisans and craftsmen also adopted
the recipe form as a means of "writing down experience" to communi-
cate instructions for producing a wide range of objects, from colors, dyes,
and varnishes to metal and glasswork.[15] Finally, often bringing together
all these spheres of knowledge, "professors of secrets" such as Leonardo
Fioravanti (1517–1588) and Girolamo Ruscelli (1500–1566)—or Alessio
Piemontese, as he is better known to readers of the genre—wrote works
that circulated across Europe in a number of vernaculars.[16] The broad
category of "recipe knowledge" usefully ties together these sets of short
instructions encoding know-how about making all sorts of things, focusing
on materials and production processes, and stemming from firsthand observation and hands-on practices.

While such practical knowledge has always been within the purview of histories of science, research on how-to or recipe knowledge has intensified in recent years, in part because of two interconnected historiographical trends in our field. The first is the call for a more inclusive definition of scientific and technological knowledge and for a focus on the multifaceted processes of knowledge production.[17] The practices of knowledge making have received particular attention, and it has become clear that vernacular practical knowledge played a central role in early science, medicine, and technology.[18] Second, in the past few decades historians of science, medicine, and technology have scrutinized and vastly extended the traditional spaces and sites for making knowledge. Alongside these studies historians have also uncovered a wide range of historical actors participating as knowers and as makers. In particular, the important role women played in various areas of science and medicine has come to the fore.[19] For example, early modern women's crucial work in the interconnected fields of pharmacy, natural history, and alchemy has been the focus of a number of studies.[20] The recovery of women's work in science has prompted scholars to analyze how gender dynamics and relations shaped knowledge-making practices and how these practices in turn influenced notions of gender.[21] As outlined below, this book builds on and extends current conversations in these important historiographical movements.

Ways of Knowing in Seventeenth-Century England

Recent studies in the history of early modern science have emphasized that within a particular cultural setting there are multiple "ways of knowing" and exploring nature.[22] Seventeenth-century England, the focus of this book, is particularly rich in this regard. This was the time of William Harvey, Robert Boyle, and John Ray and the time that saw the founding of the Royal Society and the beginnings of experimental philosophy. However, it was also the time when Hugh Plat's *The Jewell House* flew off booksellers' shelves and when Simon Forman and Richard Napier drew up astrological medical charts for scores of patients each day.[23] By surveying and analyzing household recipe books and recipes, this book extends our current narratives of early modern knowledge making.

The householders featured here might be new to historians of science and medicine, but they were firmly part of a number of overlapping conversations in early modern England. Actors such as Johanna St. John corresponded with figures like John Locke, probably read Hugh Plat's *Delight for Ladies*, and regularly consulted their brewers and distillers.[24] These different groups of knowers in seventeenth-century England, from virtuosi to London merchants to artisans to householders, shared deep connections in the way they explored and investigated the natural world.[25] As I show, not only did they often share a space—the household—for their scientific work, as well as common production processes and equipment, they also shared a sensibility in privileging empirical and experiential knowledge.

The kinds of knowledge practices I describe here sit comfortably alongside a number of epistemic schemes. Householders' zeal to collect large numbers of recipes connects their work to several knowledge-making models in early modern England. We may situate this enthusiastic gathering of "experiments," to use our actors' term, alongside other contemporary practices like such large-scale endeavors as Francis Bacon's manuscript notebooks, later published as *Sylva Sylvarum*, or the work of the London merchants Clement Draper and Hugh Plat.[26] Concurrently, they also have much in common with Baconian experimental histories, or *historia experimentalis*, where as in the *historia* tradition, practitioners sought to collect facts while abstaining from formulating conceptual frameworks or causes.[27] First promoted by Francis Bacon (1561–1626) and later by Robert Boyle, experimental histories consisted of "the collection of experiments of arts" that "exhibit, alter and prepare natural bodies and materials of things."[28] Ursula Klein and Wolfgang Lefèvre have demonstrated that experimental histories played a crucial role in histories of seventeenth- and eighteenth-century chemistry.[29] The household was, of course, also a key site for early chemical practices.

In their practical-mindedness and their focus on materiality, the householders in this book also have a great affinity with artisans and craftsmen. Scholars such as Pamela Smith have argued that the artisans' practically and materially oriented paths of investigating nature form "vernacular epistemologies."[30] These texts are thus instructions for doing and making and constitute "attempt[s] to capture in writing—perhaps to teach by modeling—the tacit, bodily knowledge of the manipulation of matter by the human hand—in other words, to capture that elusive human ability, skill." Reading a recipe, Smith tells us, "cannot be separated from trying the methods recorded in it."[31] The central chapters of this book outline in detail the various practices householders adopted to try and to test medicines, materials, and production processes, connecting them, through their knowledge practices, with contemporary artisans, craftsmen, and makers and with various kinds of practical knowledge.[32]

By showcasing householders as makers of knowledge in their own right and in studying them through their own words and writings, this book contributes to a broad and integrated narrative of science in seventeenth-century England, examining the intersections of multiple ways of knowing. I illustrate deep connections in epistemic practices among various early modern knowledge communities, arguing that they share an intense curiosity about the natural world and a common framework of multi-step, structured processes of knowledge making, with strong emphasis on experience and observation. Yet, as will become clear, I also argue that the location of these epistemic practices deeply shapes the frameworks for knowledge making. By delving into practices of everyday sociability, household provisioning, and family record keeping, we can study how one site—the household—framed early modern knowledge making.

The Early Modern Household as a Site of Science

In recent years, alongside artisanal workshops and Renaissance courts, early modern households have emerged not only as social, economic, political, and intellectual hubs but also as one of the main locations and driving forces for a range of epistemic enterprises.[33] The household was a particularly important frame for recipe-related activities. Homes were well equipped for making natural knowledge, and particularly recipe knowledge. For those seeking knowledge through reading and writing, studies, libraries, and closets offered spaces to store, consult, and take notes from print and manuscript books and to compose new tracts. Gardens, deer parks, henhouses, and other outdoor spaces presented opportunities for making botanical inquiries and observations on the life cycle of animals. For example, chapter 2 illustrates how gentlewomen like Johanna St. John used their country estates to explore the life and growth cycles of turkeys or the way plants reacted to particular types of soil and environmental conditions. Spaces like kitchens and stillrooms were well equipped for investigating the heating, melting, and distilling of substances. Contemporary prescriptive manuals and account books reveal that among the financially elite it was common to have on hand alembics, stills, mortars and pestles, furnaces, scales and weights, and much more.[34] In other words, they owned instruments and equipment that would not be at all out of place in an apothecary's workshop or an early modern chemistry laboratory.[35] Aside from this range of equipment, many of the people who worked in the spaces—brewers, kitchen maids, distillers—had the expertise and experience to use these items.[36] As I show in chapter 4, on recipe trials, by using everyday processes such as brewing, boiling, and distilling to manipulate materials, householders continually searched for new knowledge. These quotidian activities were widespread and involved a range of knowledge, skills, and expertise. It is thus not surprising that households were ideal spaces for investigating the natural world.

However, previous studies on science in the household have tended to explore how the household served as a contextual frame shaping and conditioning the scientific endeavors of well-known men of science. Steven Shapin's groundbreaking essay "The House of Experiment" argues that gentlemen's private residences were the sites of much of the experimental work of the day. Shapin offers copious examples, from the work of iatrochemist and physician Francis Mercury van Helmont (1614–1698/99) at Ragley House, the residence of the Conway family in Warwickshire, to the laboratory of chemist and physician Frederick Clodius (1629–1702) in his father-in-law Samuel Hartlib's kitchen in Charing Cross, to Sir Kenelm Digby's house and laboratory at Covent Garden.[37] A number of studies have followed by bringing to light the rich practices that householders, both male and female, engaged in within the walls of early modern homes. For the most part, though, these studies focus on well-known men of science such as John Dee (1527–1608), Ulisse Aldrovandi (1522–1605), Robert Boyle (1627–1691), and René-Antoine Ferchault de Réaumur (1683–1757) and illustrate how their domestic situations played an integral part in their scientific endeavors.[38] Not only did they conduct experiments within the home, but household members (servants, wives, and daughters) played an important part in creating knowledge. These "invisible technicians," to use Shapin's term, did a number of tasks, from managing air pumps to observing and recording experimental outcomes to sketching insects and botanical specimens to managing the smooth running of the household to enabling scientific exchange.[39]

This book both builds on these past studies and departs from them. Looking beyond the domestic situations of learned men, it investigates science in the home from the ground up, reconstructing knowledge-making activities layer by layer in a wide range of households. Driven by curiosity and the very real need for trustworthy information, early modern men and women engaged with recipe knowledge in a number of ways. They gathered and assessed knowledge on paper, comparing the newly gathered recipes with familiar and trusted know-how from conversations with books, experts, and family and friends. They also made repeated "recipe trials" for several reasons, including testing their efficacy and the effects of materia medica on human and nonhuman bodies, observing the behavior of particular substances under heat or other external stimulus, and trying out body techniques and ways of encapsulating gestural knowledge. Through these engagements householders gained deeper understanding of the human body, of natural and man-made processes, and of everyday materials. Their recipe books are the records of these trials and witnesses to observations and investigations of natural particulars and to efforts to accumulate and record natural knowledge.[40] By taking household recipe knowledge as a starting point, this book uncovers and analyzes these home-based practices as knowledge making in themselves.

Concurrently, while acknowledging that it was often gentlewomen and gentlemen who wrote down household recipes, I show that they were aided by teams of men and women who had hands-on experience and expertise in a range of practical tasks. This book investigates their shared practices. By turning our lens on the knowledge making of household collectives, I seek to reposition men and women hitherto referred to as invisible technicians as makers, writers, and transmitters of natural knowledge rather than as merely auxiliaries to well-known men of science. Exploring the different roles adopted by husband and wife and by master and servant, I also examine the gender and class dynamics inherent in group-based knowledge making. By recovering householders' practices I widen the cast of historical actors participating in and contributing to knowledge and extend our narratives of early modern natural inquiry, unveiling the early modern household as a vibrant site for making everyday knowledge.

Recipes, Families, and Households

The "household," both as a site of knowledge production and as a collective of knowledge producers, provides a central analytical and conceptual framework for this book. The physical space and architecture of the early modern household crucially framed recipe-related activities. In particular, the sprawling estates of the English gentry's country houses provided opportunities for householders to pursue knowledge about nature. Parlors and dining rooms hosted conversations about making cough remedies, distilling face waters, and preserving fruit. Kitchens, stillrooms, and brew houses accommodated the trials and tests of newly proffered recipes and the often time-consuming production of medicines and foods. Kitchen gardens, henhouses, and stables were places where householders indulged their natural historical interests as they raised turkeys for Christmas or cultivated prized French melons.

I also use "household" to refer to a group of historical actors connected in overlapping ways and living and working together as a collective.[41] Masters, mistresses, husbands, wives, brothers, sisters, sons, daughters, stewards, housekeepers, gardeners, and dairymaids all make an appearance in our story, and for the most part they worked collaboratively. We encounter a diverse range of collaborations from husband-and-wife teams like Henry (d. 1665) and Mary (d. 1649) Fairfax to father-and-daughter pairs like Peter (d. 1660) and Elinor (d. 1729) Temple. Additionally, the gentlemen and gentlewomen in these cases were often aided by a host of domestic servants. Some of these characters, such as Johanna's steward Thomas Hardyman (d. 1732) and his wife, Elizabeth, come to our attention, but many others leave little trace.[42] Recipe-oriented knowledge practices were often formed within these complex webs of social relationships.

By emphasizing collective engagements, this book views household knowledge making as a series of collaborative household-wide endeavors, filled with continuing gender and class-related negotiations.[43] It thus extends recent research on early modern English recipe books, which has tended to highlight women's medical practices and privilege discussions of women's manuscript writings.[44] Traditional social, class, and gender hierarchies, of course, shape the practices I showcase. But the case studies also demonstrate that notions of authority, obligation, and responsibility shifted according to specific situations. Then as now, families tended to be governed by their own relationship dynamics that were shaped by factors from gender to birth order to social role. Since many of the recipe books were created by generations of family members over long periods, the relationship dynamics between knowledge producers and users must necessarily also shift and change.

The location of these practices within early modern homes brings with it, I contend, particular frameworks for knowledge making. Here *Recipes and Everyday Knowledge* makes three primary claims. First, making recipe knowledge was very much part of household management and large-scale planning for provisions. Householders were acutely aware of seasonality, and producing home-based medicines went hand in hand with scheduling and preparing foods such as butter, perfectly ripened cheese, fattened poultry, and well-cured bacon. The juxtaposition of food and medicine in so many recipe books resulted from the multifaceted role taken on by household managers, who moved seamlessly among this colorful array of endeavors. Consequently the collecting and gathering of practical know-how and the practices of making these recipes must be situated within the bounds of early modern household management.

Second, the book demonstrates that home-based knowledge gathering was built on existing family and social relationships. Not only did householders seek know-how within established networks, they also used long-standing knowledge about the general trustworthiness and expertise of donors to gauge the proffered recipes. Consequently, the creation of household recipe books has as much to do with recording a family's connections as with gathering recipe knowledge. Additionally, like medicines and foods, recipes also played a key role in economies of patronage and gift exchange.[45] Thus household recipe books not only were active maps of a family's social network but were in effect ledgers or account books recording its social obligations and credits.[46]

Finally, recipe collections took on yet another role—as records of family activities and archives of family histories. In many families, members— male and female and cross-generation—worked together to gather, test, and record recipe knowledge. The resulting notebooks document these collaborations and collective practices. Recipes were often preserved alongside other family records such as legal and financial documents and genealogies. It is therefore not at all surprising that recipe collections were considered family heirlooms and passed on from one generation to the next. In fact, family provided the main vertical axis through which recipe knowledge traveled, and many recipe books remained within the bricks and mortar of the family seat for hundreds of years. Recipes were one way families remembered everyday experiences and recorded their family histories and identities.[47] The social and cultural importance of these notebooks goes a long way to explain the abundance of surviving examples, and this dual framing—kinship and sociability on one hand and household management on the other—influenced not only the recording of recipe knowledge but also, significantly, acts of testing and trying.

The Household Recipe Archive: Sources and Methodologies

On 7 March 1703, a little shy of one year before her death, eighty-four-year-old Lady Johanna St. John wrote out her will in her own hand. St.

John's will contains the usual list of personal bequests—she leaves her Bible to her eldest son, various items of furniture, paintings, and silver to her daughters and granddaughters, and small gifts of cash, clothes, and linens to loyal old servants. Two handwritten recipe books were listed alongside these gifts. The first was a book of recipes for cookery and preserves to be left, "according to [her] promise," to her granddaughter Lady Johanna Soame. The second, given to her daughter Lady Anne Cholmondeley (d. 1742), was her "great Receit Book" filled with instructions for making a range of medical remedies.[48]

The "great receipt book" listed in St. John's will is a quarto-sized leather-bound volume with "I. S." stamped in gold on the cover. Alphabetically organized, the book houses hundreds of medical recipes. Under the letter *A*, we find recipes for ague and aches.[49] Remedies for "kankers" in a woman's breast, back pain, bone aches, nosebleeds, poisonous bites, and bruises fill the section titled *B*.[50] The *C* section brings instructions for making cordial water and cures for consumption, coughs and colds, colic, worms in children, convulsive fits and cramps, and so on.[51] The rest covers all kinds of ailments, illnesses, and discomforts of the body, and the book is a treasure chest of potential cures, all collected "just in case" a family or household member falls victim to ill health.[52] Perhaps reflecting the St. Johns' ever-growing family (the couple were the parents of thirteen children), the book also contains a wealth of recipes to promote fertility, to discourage miscarriages, and to ensure an easy delivery. All these were filed under *W* alongside cures for worms, wounds, wind in the stomach, and making distilled waters. While St. John clearly had a separate book for her cookery recipes, veterinary recipes such as ones titled "For A Bulock that pisseth Blood" and "For the Itch in man or woman or mang in a dog" are recorded alongside medicines intended for members of her household.[53]

Johanna St. John's recipe book is representative of other household recipe books in early modern England. The books are curious objects: many are worn, clearly heavily used and read, and written and annotated by many hands. Many were open texts containing a wealth of miscellaneous information, and flipping through one of these one often finds poems, shopping lists, letters, and family records interspersed with medical, culinary, and household recipes.[54] Like St. John's gifts to her daughter and granddaughter, a good proportion of these books also passed through the hands of multiple owners, compiled and used by generations of householders.[55]

By the time Johanna St. John created her book and willed it to her family, recipes and recipe collections, as a textual genre, had been in use and circulating for centuries. Gathering medical information in short, concise recipes stretches back to the ancient world, and from then on it functioned as a central way medical and practical knowledge was transferred and circulated in cultures across the globe.[56] Within the English context, recipes have long been the mainstay of vernacular literature, and some of the earliest vernacular medical writings were in fact recipe collections, including Bald's *Leechbook*,[57] Middle English translations of Gilbertus Anglicus's pharmaceutical writings, the *Liber de Diversis Medicinis*, and the *Tabula Medicine*—a set of fifteenth-century manuscripts described by Peter Jones as an "evolving encyclopedia" to emphasize its ever-changing nature and the miscellaneous quality of the medical information it contained.[58]

The long seventeenth century in England, the focus of this book, saw an unprecedented production of manuscript and printed recipe books. Handwritten books of medical, culinary, and practical recipes are common in today's archives and libraries, and hundreds of examples have survived. Alongside this flourishing manuscript culture, the period also witnessed a boom in printed recipe collections. Contemporary printers/publishers and booksellers enticed readers with a range of titles, from books marketed as the treasuries of noblewomen, like *A Choice Manual* (1653) and *The Queens Closet Opened* (1655), to those connected with contemporary medical practitioners, such as Thomas Brugis's *The Marrow of Physicke, or A Learned Discourse of the Severall Parts of a Mans Body* (1640), to advertisements like Francis Dickenson's *A Pretious Treasury of Twenty Rare Secrets* (1649).[59] In fact, between 1600 and 1700 the London book producers issued over 200 medical recipe titles, consisting of 60 new titles and about 170 reprints, making them one of the most popular genres of vernacular medical texts in early modern England.[60]

While recipe books are common across premodern Europe, most of the examples we encounter, such as the hundreds of secrets in *I secreti del reverend donno Alessio Piemontese* (Venice, 1555) or Hugh Plat's *The Jewell House of Art and Nature* (1594), are formal treatments in print representing already codified knowledge, with little hint of the underlying theoretical framework or epistemic practices.[61] Handwritten household recipe books, on the other hand, were malleable texts open to correction and extension. Often filled with users' annotations and cross-outs, they record the multiple instances where householders tested and amended the instructions. We might think of them as precursors to modern laboratory books or "research notebooks," which historians of science have argued reveal "science in the making."[62] Like modern laboratory books, they record the nitty-gritty of everyday experimentation and are ideal entries to the early modern world of making and knowing.

Crucially, household recipe books like these tend to lack contextual information in multiple senses. Many offer only recipe after recipe, with few additional instructions or clear articulations of historical actors' intentions, ideas, and goals. Second, most manuscript recipe books now housed in libraries or archives have survived with little or no accompanying biographical information. Consequently we have scant knowledge about the creators, readers, and users of these texts, and it is difficult to locate the historical actors and spaces that shaped their creation. The patchy survival of accompanying personal archives and papers is in part due to the social and economic contexts within which the texts were produced. Many of the historical actors in this book were wealthy landowners—members of Parliament, minor gentry, prominent local families—who had the leisure to engage in natural inquiry and the education to record their endeavors in writing. However, the survival of personal and domestic papers is dependent on the whims of later generations.[63] The challenge these sources pose prompts us historians to turn to new methods and ways of thinking about these texts. Two central approaches underpin this book: studying the "household recipe archive" and giving attention to the materiality of texts.

The sheer number of surviving texts invites us to examine them as a coherent group. While each chapter focuses on a few case studies, the project is informed by a large-scale survey of manuscript *and* printed recipe collections produced in seventeenth-century England. I aim to tease out the idiosyncrasies of individual households and families as well as to make statements about household recipe knowledge in general. Arguably, each recipe collection is tied to an individual household and presents a series of unique practices; taken together and surveyed as a whole, however, the hundreds of household recipe collections are connected through a series of practices common to early modern households. I also posit that there are strong overlaps in how print and manuscript collections were read and used. Marginal annotations in both printed and manuscript examples demonstrate similar patterns of trying and testing recipe knowledge. Paratextual materials in both media suggest a lively dialogue between recipes in the two, which together represent contemporary recipe knowledge in a written form. Together they form what we might call a household recipe archive.

The household recipe archive at the heart of this book consists of about 260 examples of English manuscript recipe books and over 200 printed titles issued from 1600 to 1700.[64] The archive under scrutiny is thus bound by geographical and temporal limitations. In terms of the manuscript samples, unlike other studies of recipe collections, which largely focus on texts compiled by or geared toward women, my survey was limited not by gender but rather by geographical and chronological criteria. Because of this, this study reveals the rich collaborations between husbands and wives, fathers and daughters, and mistresses and servants. While acknowledging that writing, collecting, and exchanging household recipes occurred across Europe, I believe the abundance of surviving manuscript examples makes the English case particularly rich for study.[65] Most of the manuscript recipe texts examined in this book were created during the seventeenth century and the first few decades of the eighteenth, a short window in the history of their making and use in England.[66] Certainly, as sketched above, there was a flourishing tradition of recipe writing in English stretching back to the medieval period, and our archives also contain significant numbers of sixteenth-century examples, including the notebooks of figures like Clement Draper and Hugh Plat, who have been the focus of modern studies.[67] However, the long seventeenth century saw the production of such texts flourish in well-to-do households. Examples from this period carry characteristic features: ownership notes and author citation tags, information retrieval devices, and clear mise-en-page.[68]

Second, I pay particular attention to the materiality of household recipe books as objects and as texts.[69] In this I follow the lead of recent books by Lauren Kassell and Deborah Harkness, whose detailed and nuanced readings of Simon Forman's casebooks and Clement Draper's prison notebooks bring quotidian medicine and science to life.[70] Many household recipe books are heavily used, worn, and marked-up objects inviting us to venture into the margins and between the lines to uncover evidence of practice. Initials embossed on their leather-bound covers might betray past owners and the objects' life stories; cross-outs and selection marks give us a glimpse into the complex knowledge-making practices householders adopted. My analysis uses methods developed by historians of reading and note taking; however, since the nature of recipe knowledge demands that each recipe be continually put into action, the marks surrounding the text in household recipe collections, whether in manuscript or in print, thus reflect both reading *and* making.[71] The focus on users' marks is particularly central to the arguments presented in chapter 3, "Collecting Recipes Step-by-Step," and chapter 4, "Recipe Trials in the Early Modern Household," where close, detailed textual examinations present opportunities to identify schemes for codifying knowledge and ways of assessing and testing recipes.

Plan of the Book

Recipes and Everyday Knowledge opens with "Making Recipe Books in Early Modern England." Centered on the notebooks of Archdale Palmer (1610–1673) and the Somerset-based Bennett family, this chapter presents a general overview of patterns of recipe collecting: adopting "starter" collections and gathering single recipes. As I will show, many families cultivated local social relationships and used them to extend their treasuries of recipes. The chapter situates the gathering and writing down of recipe knowledge alongside a range of social practices, from forming alliances to giving gifts. I thus demonstrate that manuscript recipe collections had a dual role: on one hand as repositories of recipe knowledge and on the other as ledgers recording social ties, credits, and debts. Social structures, local networks, and alliances shaped recipe knowledge in crucial ways, from information access to record keeping to practices of trying and testing.

Chapter 2, "Managing Health and Household from Afar," places collecting and codifying recipe knowledge within the frame of household management and planning in two ways. First, I explore long-term provisioning within early modern households and situate medicine making alongside the production of foods and other processed goods. I contend that the juxtaposition of food and medicine in so many recipe books was a product of the multifaceted role taken on by early modern housewives and household managers, who moved seamlessly across a wide range of activities. Our narratives of household medicine and science thus must also take this more holistic approach. Second, this chapter connects the daily producing of food and medicine with larger economies of patronage and gift exchange. Looking after the "health" of a household went beyond offering cures for coughs and colds to include vigilant monitoring and maintenance of the household's social and financial health.

Assessing knowledge and trying cures form the main focus of chapter 3, "Collecting Recipes Step-by-Step," and chapter 4, "Recipe Trials in the Early Modern Household." Chapter 3 contends that contemporary men and women evaluated potential know-how through multiple steps before assimilating new medical recipes into their treasure chest of home-made cures. Within this scheme of codifying knowledge, making and try-ing recipes played a central role. I explore this process in three settings. First, I use the letters of Edward Conway (1594–1655) and Edward Harley (1689–1741) to outline how two historical actors conveyed this process of codifying knowledge. Second, I turn to a series of three notebooks created by Sir Peter Temple (d. 1660) to demonstrate how recipe compilers copied and recopied recipes from notebook to notebook as the knowledge was assessed, tested, and evaluated. Finally, using an analysis of users' annota-tions and marks in several household books, I show that the schemes out-lined in the Conway/Harley and Temple case studies were widely adopted.

Opening with Sir Edward Dering's busy summer of trying and test-ing drugs, chapter 4 explores recipe trials. Building on the arguments pre-sented in earlier chapters, I argue that the domestic and family setting brought a particular set of considerations in determining expertise, au-thority, and value. Investigations into "making trials" on medical recipes in the household thus offer us insight into assessment and testing in a new spatial and social context and give us a rare glimpse of practices on the ground, outside academic institutions. Within this context, making goes hand in hand with testing and writing down. As householders produced medicines and foods from written recipes, they were continually testing them for efficacy, for the way materials behaved in particular contexts and environments, and to see how they reacted with the human body.

Chapter 5, "Writing the Family Archive," explores the multitude of ways householders used collecting recipes and creating recipe collections to construct and write their own family histories. I argue that gathering recipes and creating recipe collections constituted one aspect of what we might call the "paperwork of kinship." Early modern householders, I show, wrote down, collated, and preserved all kinds of paperwork concerning the social and economic holdings of the household, from land deeds to rent accounts to lists of births and deaths. Working together, these docu-ments not only sketch out a social and economic history of a family but also construct its very identity. Recipes and recipe books, I contend, were a crucial part of this paperwork, and it is in no small part due to this role that so many examples survive in the archives.

The book concludes with a chapter titled "Recipes for Sale: Intersec-tions between Manuscript and Print Cultures." By the 1650s, the London booksellers' shelves were packed with printed recipe collections. With titles like *A Choice Manual of Rare and Select Secrets* (1653), *The Marrow of Physicke, or A Learned Discourse of the Severall Parts of a Mans Body* (1640), or simply *A Pretious Treasury*, these books offered English readers hundreds upon hundreds of recipes. This final chapter explores the intersections, commonalities, and differences between the manuscript and printed recipe collections of the period. Through analyzing traces of reading and writing, I emphasize the significant crossovers between these two media. The media of communication, I contend, crucially shaped the kinds of knowledge transferred.

CHAPTER ONE

Making Recipe Books in Early Modern England

Material Practices and the Social Production of Knowledge

Archdale Palmer (1610–1673), gentleman landowner, high sheriff, and lord lieutenant of Leicestershire, had a habit of collecting recipes.[1] On 27 November 1658, Palmer, then in his late forties, took his first step toward creating his collection by purchasing a blank notebook for the sum of sixpence. To mark the occasion he twice wrote on the flyleaf, "Arch: Palmer his book" and the date he bought it. A little over a month later, on 3 January 1658/59, Palmer entered his first tidbit of medical information—a recipe "for running gowte" collected from "cozen" Mary Williams. For the next fourteen years Palmer maintained a healthy interest in recipes and an active collecting habit. He continued to add to this notebook regularly, gathering new recipes every few months. On 8 June 1672, a little over a year before his death on 6 August 1673 at age sixty-three, Palmer wrote his last entry. By then he had amassed more than 150 pieces of medical, culinary, household, and veterinary information from his family, friends, and chance acquaintances.

Almost twenty years after Palmer collected his last recipe, another family from another county began a similar project. The Whitney Collection of Cookbooks Manuscript 9 is a household recipe collection written in what was once a commonplace book. The notebook contains nearly three hundred recipes, scattered among sections of philosophical and religious writings.[2] In contrast to Archdale Palmer's notebook, where authorship and agency are clearly stated, the compilers of the Bennetts' notebook were more reticent in asserting their role as creators of the work. A recipe likely written in 1707 or 1708, titled "for my son Samuel Bennetts lame Knee William Coleman of Brewham used when he was neere 13 years of Age" connects the collection with Philip Bennett the younger (d. 1725) and Anne Strode Bennett (d. 1735) of Somerset.[3] Starting in 1694, for nearly four decades the Bennetts worked together to gather recipes from their family, friends, and neighbors and recorded them in their notebook.

Like the scores of other household recipe books surviving from the period, the Palmer and Bennett books contain a miscellany. Readers find instructions for making remedies to alleviate a wide range of general ailments such as scalds and burns, rickets, giddiness of the head, gout, ague, rupture, nosebleeds, and toothaches. Archdale Palmer and the Bennetts cast a wide net when seeking out recipes, and their notebooks present an assortment of information taken from oral, manuscript, and printed sources described as "Perkins' almanac" or "Dr Micklethywat" or "sonne Wm Palmers wife Martha."[4]

We know a great deal about the collecting habits of both sets of compilers because they took unusual care in recording the time, location, and source of recipe donations. The two notebooks were organized chronologically by date of entry, and this journal-like arrangement enables us to situate the practical know-how needs and the place of recipe collecting within the everyday lives of the Palmers and the Bennetts.[5] Reading between the lines, it is clear that though adding to their recipe books might not have occurred daily, for both families seeking out and collecting new recipe knowledge was constant and sustained for many years. Filled with bountiful details, the collections also offer tantalizing hints to Palmer's and the Bennetts' social and knowledge networks and to social occasions, spaces, and actors involved in the exchange of recipes. Additionally, they reveal contemporary assessment criteria for collating practical knowledge and the various processes used to construct treasuries of household know-how. Together these two books provide a unique glimpse into both processes of recipe knowledge collection and the social worlds of early modern England.

Guided by these two recipe collections, this chapter explores the multiple paths by which householders encountered recipe knowledge and the various ways they recorded it. The chapter opens with a general overview of patterns of recipe collecting: adopting "starter" collections and gathering single recipes. The rest focuses on four main themes: the importance of family; sociability and knowledge making; recipes and gift econo-mies; and the social and cultural function of recipe exchange. Overall I argue that within early modern households, gathering recipe knowledge was shaped by family and social networks, economies of social credit and debts, and systems of gift and information exchange.

Getting Started

In 1694, when Philip and Anne Bennett began their recipe book project, they turned for information to their neighbor, the widowed Florence Mompesson. Thomas (d. before 1693) and Florence Mompesson (d. 1698) and their elder daughter, also named Florence (d. c. 1709), lived across the river Brue at a nearby farm known as North Court or Batts. The entries in the Bennett notebook suggest friendly relations between the two house-holds.[6] Before her death, the elder Florence Mompesson contributed in several ways to the Bennetts' notebook of recipe knowledge. First, when Philip and Anne began their recipe book in 1694, they copied a long series of culinary recipes from Mompesson's collection to form their own "starter collection"—a practice I discuss further below. As Philip and Anne began gathering single recipes to build on this "starter" col-lection, they again turned to Mompesson. She obliged on two occasions and offered them instructions on making ointments, oils, and remedies to ease burns and rickets.[7] A number of these had themselves been given to Mompesson by others. The ointment recipe, for example, came with the endorsement that Mompesson's Aunt Drew of Exeter had "cured many of the cancer at the first coming," and the recipe for the oil of charity came with the tag "This madam Moore did much good with."[8] By sharing her medical and culinary knowledge with the Bennetts, Mompesson opened access to her own network of family and social contacts. This group of "Mompesson" recipes in the Bennett family notebook highlights both the common pattern of recipe collecting and the important role of sociability in making recipe knowledge.

When Philip and Anne Bennett chose to adopt a section of the Mompesson recipe book as the first step toward their own collection, they were joining in a well-established practice. Whether they obtained them from family members or from friends and acquaintances, many rec-ipe compilers began their books with "starter" collections. These take the form of a large section of copied text written in a uniform hand, occurring at the beginning of the manuscript. The notebook is then filled up with a substantial number of additional recipes in a different hand or hands. This combination of a group of recipes clearly transcribed from another collection and additional recipes gathered singly is one of the most common constructions. Many of these starter portions are written out in a neat scribal hand, in contrast to the less practiced hands used to fill in the blank pages with additional recipes, and they are often well organized, with an index or a table of contents. In fact, with many of the collections, the method of organization disintegrates only after the additional recipes are put in. Many starter collections are copied leaving blank pages for later additions; some were written only on the recto pages, leaving the verso pages to be filled by subsequent owners of the notebook. If indexes and tables of contents accompany the starter portion of the collection, space is left within them for additional entries.

Aside from the Bennett family book, another good example of a "starter" collection is the notebook owned by Anne Brumwich, Rhoda Hussey Fairfax, Ursula Fairfax, and Dorothy Cartwright.[9] The first folio of the manuscript bears the following inscription: "Mris Anne Brumwich her Booke of Receipts or Medicines for severall sores and other Infirmities." The hand that wrote this inscription also added a set of recipes organized alphabetically by ailment. Thus the Brumwich hand begins with recipes for the ague and continues with recipes for consumption, cough, and so on. At the end of each section, space is left for additional recipes. An index was constructed at the end of the volume, where space was again left for additions. The recipes written in the Brumwich hand are not attributed to any particular author. The uniformity and organization of these recipes suggest they were copied to order out of Anne Brumwich's book—all the recipes were by Anne Brumwich, hence there was no need to reiterate authorship. Once the starter collection came into the hands of Rhoda Hussey, soon to be Rhoda Fairfax, she began to fill in the empty pages with her own recipes for the same ailments and updated the index. The subsequent owners of the volume, Rhoda's daughter Ursula and granddaughter Dorothy Cartwright, also added their own recipes. The book as it exists now has almost alternating sections of the Brumwich hand and the later hands. The recipes added by the later hands bear the signs of compilation, and most note their authorship or origin. Rhoda Fairfax sometimes recorded the date when she obtained a particular recipe. Examples include "a pill commended for mee R. F. by Doctor Catlin, 1653 when I was very ill"; "The Yellow Salve—Lady Hussey's

29 September 1684," and a recipe on a loose piece of paper titled "this was sent from London for the payne in my shoulder . . . R. F. the 29th July 1682."[10] It appears that the copy of Anne Brumwich's collection provided Rhoda Fairfax with a base of medical knowledge and information that she built on and personalized during her lifetime, then passed on to her daughter.

Not surprisingly, these starter portions are common in early modern recipe notebooks. Many householders elected to copy or adopt a group of already assembled recipes as the basis of their own collection. We might think of these as initial assemblages of household knowledge to give compilers a head start. The origin of these starter portions varied. Some, like the Bennetts, obtained their starter collections from neighbors or friends. Occasionally, as with those of the Temple and Bourne families we will encounter in later chapters, a family member might take the time to copy out the starter portion. In other cases compilers might themselves take the initiative to create these books, either by doing the copying themselves or by commissioning a scribe, as when Anne Fanshawe hired Joseph Averie to copy the collection of her mother, Margaret Harrison.[11]

The presence of starter collections indicates that both the producer/donor and the receiver/future compiler saw a need for a ready-made set of recipes. Yet this need for general information was paired with a desire to personalize these collections and adapt them to one's own requirements—hence the blank spaces left in each collection and the subsequent evaluation of the information they contained. Household recipe collections were by nature ever-expanding books of knowledge that changed according to the needs of the current owners.

Whether armed with the starter collection or working from scratch with a blank notebook, men and women like Palmer and the Bennetts actively gathered single recipes or groups of recipes through various avenues. As I outline in the following sections, family, friends, and neighbors provided householders with an abundance of recipe knowledge. Yet family and social networks did much more than just supply new information; as I discuss below, they also shaped the very making of recipe knowledge.

Working Together as a Family

Family played a central role in collecting, exchanging, and recording recipe knowledge. Family members such as husbands and wives, fathers and sons, and mothers and daughters collaborated on their collections. Recipe gathering was often directed by the sickness and needs of individual family members and other household dependents. Additionally, recipe compilers often reached out to extended family and kin for new recipes. In this section I examine each of these areas in turn—collaborative compiling, collecting to cure, and calling on family and kin—to argue that making household recipe knowledge was a family affair.

Collaborative Compiling

When Philip and Anne Bennett decided to start writing down recipes in an old commonplace book, they were a long-married couple with a large and well-established family. Recipe collecting was a joint project for them. A recipe for "the collick," gathered on 3 May 1719, comes with the endorsement that "my wife took it," revealing Philip Bennett's active role in the project.[12] Entries made after Philip's death in 1725 show that Anne also contributed to the family book. During this period it was not unusual for male and female family members to take collective responsibility for the health of the household. Contemporary correspondence reveals that fathers and husbands often wrote about the health, ailments, and subsequent treatment of their wives, children, and dependents, and at times took care to produce medicines—either with their own hands or, more likely in elite families, by instructing their servants.[13] The actions of Philip and Anne Bennett make it clear that men's interest in the family's health extended to another kind of household medical activity: collecting and recording recipes.

Philip and Anne were, of course, not the only husband and wife to work together on a recipe book. The recipe archive contains many such instances. One particularly clear example is the collection now known as *Arcana Fairfaxiana*, created through the joint efforts of the Reverend Henry Fairfax and his wife, Mary Cholmeley Fairfax. Like many of the collections described above, the *Arcana Fairfaxiana* begins with a starter collection, probably created before Mary wedded Henry in 1627.[14] This section is written in a clear hand and lists a series of medical recipes by ailment, followed by several recipes written in abbreviated Latin, suggesting that they were taken either from an apothecary's formulary or from a physician's notes.[15] In this section of the book, if a recipe spanned more than one page, catchwords were used, suggesting that the text was first copied and then bound. This same hand also wrote a note on "Miss Bar-

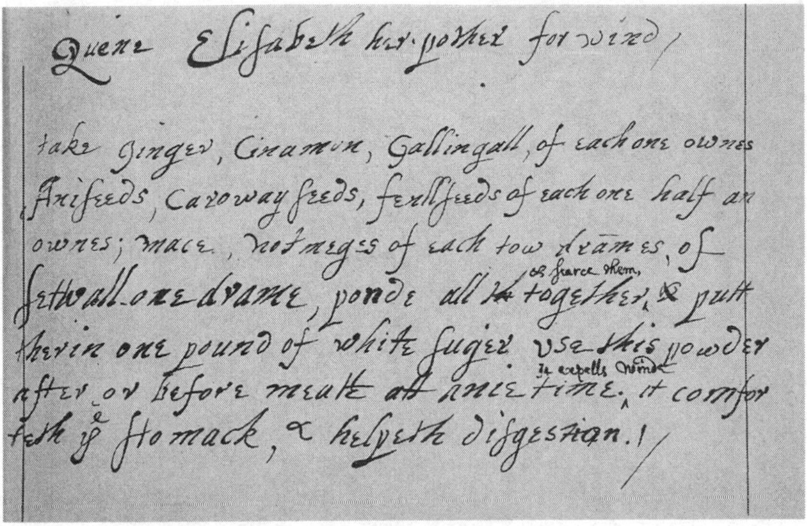

FIGURE 1.1. *Arcana Fairfaxiana Manuscripta*, 63.

bara's" lessons on the virginal and listed late sixteenth-century musicians on the back flyleaf.

Once married, Henry and Mary began to fill in their recipe book together, but Henry seems to have taken the lead. His neat, precise hand is dominant throughout the book. He inscribed his name on the first folio of the volume, wrote out and translated a Latin epigram, compiled the indexes in the front (later crossed out) and back of the volume, and added information on weights and measures.[16] He added a substantial number of single recipes on the blank pages of the notebook and also inserted several among information dealing with similar ailments. For example, he added a recipe "For a Sore Breast" under a recipe "to eale the greate heate in the brests of women or in the privy members of men."[17]

Mary Cholmeley Fairfax also contributed to the volume and wrote in many recipes interleaved with those copied by Henry. One entry, seen in figure 1.1, is particularly interesting. The recipe is titled "Quene Elisabeth her pother for wind." The recipe is entirely in Mary's hand, but it bears corrections and additions in Henry's. Mary's version instructed the reader to "ponde all [the ingredients] together"; above this Henry added "& searce them." Even though the recipe is titled "for wind," Henry felt compelled to reiterate that "it expells winde" in Mary's list of additional virtues.[18] This is not the only recipe written by Mary and annotated by her husband.

In a recipe to make "the green oyntment," Mary wrote that one should take "red sage and rewe of ech a quart," and Henry added above, "a pound or [a quart]." The quantity of two other herbs used in the recipe, bay leaves and wormwood, was given in pounds, and thus Henry may have added the extra measurement to ensure consistency.[19] While there are scant written traces that Mary returned the favor by modifying Henry's entries, another family member, yet to be identified, took the liberty of doing so. Written at the end of the book is a recipe by Henry offering instructions for making a consumption plaster. While Henry's original instructions are frustratingly vague on the amount of ingredients required, a second hand clarified this by adding precise measurements and specific instructions for application—the "plaister is to be laid to the spoone of the stomack & cut to the bredth of a hand."[20] Aside from these contemporary annotations, other family members also entered this collaborative project at various times.

In later years Henry and Mary's younger son, Brian Fairfax, also took an interest in the household recipe book and wrote several entries, including a drink for the plague, a copy of the instructions to make Dr. Chamber's water, and a series of miscellaneous recipes.[21] As it stands, the *Arcana Fairfaxiana* offers clear traces of how generations of the extended Fairfax family collectively worked on their book of recipe knowledge over long time spans by entering information into the volume and by actively commenting on each other's entries. Thus household recipe books like this one are open-ended projects growing alongside the family.

These family-wide collaborations muddy the waters when it comes to gendering knowledge production. One example is Nicholas Blundell's (1699–1737) "great diurnal" where he notes down instances of recipe exchange or his own work on the family recipe book. Although Blundell refers to the family recipe book as "my Wives book of Cookery," he spent a day making an index for it, hired a scribe, Edward Howard, to write out recipes in the book, and recorded the dates of recipe donations. As Sara Pennell argues, the involvement of male members of the household "does not . . . render manuscript recipe collections invalid as a means through which to study female knowledge formation."[22] However, the collaborative aspect of recipe compiling does encourage us to acknowledge nuances in the gendering of recipe knowledge. For some families, at least, gathering recipes and recording them in notebooks was shared by husbands and wives. Responsibility for adding to the family's stock of practical knowledge flitted between the two. Undoubtedly the power dynamics between family members in these acts shifted according to individual interests, points in the life cycle, and access to resources. Acknowledging the different roles various family members adopted lets us gain a clearer picture of the making of household recipe knowledge, one characterized by collective and collaborative endeavors.

Collecting to Cure

In 1699 Samuel Bennett caught a cold. In response, a member of his family, perhaps his mother Anne, applied to Mrs. Marshall of nearby Bruton for a remedy. Just a few months later, Samuel suffered a nosebleed. His concerned mother must have mentioned his condition to kinswoman Katherine Pelham, since she later wrote in the entry titled "for to stop Bleeding of the nose cousin Katherine pelham 1700 the 17th of December for Samuel if he should bleed again."[23] This type of collecting practice— the targeted gathering of health-related knowledge to solve a problem at hand—is evident throughout the Bennett and Palmer family books.

Reading between recipes, we get some sense of a few family members' states of health. Certainly Philip and Anne Bennett were offered advice on dealing with their own bodily ailments. A remedy collected from "Goody Colling" on 5 September 1702, geared toward a tertian ague, involved boiling rue in some brandy. As the fever fit approached, the family member was instructed to apply the brandy as hot as can be suffered on the patient's wrists. As the "cold fit" approached, the patient was to drink some of the brandy mixed with water. This simple remedy was to be repeated until the patient recovered. Either Philip or Anne ended this entry with the powerful endorsement of "this cured mine."[24] In another instance, during a family visit with "Sister Shute" in Kilmersdon in June 1715, either Philip or Anne was struck by a violent cold, which was alleviated by a medicine their hosts offered. The recipe for this remedy, "for a violent cold and exceeding cough," was subsequently entered in the family book with the personal endorsement that the medicine had done much good on that occasion.[25] In this case writing down the cold and cough remedy served both as know-how for future bouts of sickness and as a record of their personal suffering.

One of the daughters, Katherine, seems to have been sickly. In April 1697 a recipe was collected from Florence Mompesson, since Katherine was feeling "so weeke and colourless."[26] She was again feeling weak in January 1709 and was simply described as "sick" in August 1713 when the

family collected two recipes in response to her sickness.[27] The first recipe
was collected from Cousin Margaret Trowbridge on 25 August, and four
days later a second recipe was collected from Dr. Rowland Cotton.[28] This
pattern suggests that when Katherine was ill the family first looked within
itself for a remedy and, when that did not perform as desired, they paid
a medical practitioner.[29] The final ailment Katherine suffered was an eye
affliction that bothered her in September 1718. The family recorded five
recipes for eye ailments in that month, including a recipe "for a bruise
or a stickes hurting an eye, Hannah Draper for my daughter Katherine,"
dated 23 September. On the same day they wrote down two more recipes "for the same." These were taken from a "Madam Madox" and a "Mr
John Albin of Brewham."[30] Further collected recipes suggest that Katherine's eye condition did not improve from using the recipes collected that
day in September.[31] That the final recipe did not come in until nearly four
months later also signifies the length of Katherine's suffering.

Like the Bennetts' collection, Archdale Palmer's notebook contains
recipes used in specific bouts of sickness. Archdale's wife, Martha, suffered from melancholy in the spring of 1662, and Archdale found a recipe from a "Pagett" to alleviate her symptoms.[32] In another instance, on
4 June 1668, Palmer wrote down a series of recipes including a cordial
julep, a cordial, and a drink for dropsy. These were labeled as for M (presumably Martha) Palmer and given by Phillippe Launder. The drink for
the dropsy receives further endorsement: Palmer notes that it was Major
Babbington's.[33] In addition to these hints about Martha Palmer's health
over the years, the Palmer family book also provides an interesting picture of community reactions to the plague. In early summer 1665, when
the plague had again broken out in London, no fewer than eight plague
cures were written into the book.[34] Many of the plague cures of the day
came with two sets of instructions—one for prevention and another for
those already infected. For the most part, the infected patient would need
to take more of the cure than for prevention, and in a few cases extra ingredients were added. The remedies the Palmer family gathered are fairly
representative of plague medicaments of the time. The ingredients listed,
such as mithridate, treacle (whether London or Venice), rue (also known
as herb of grace), and vinegar were all well-known and often-used remedies for the plague.[35] Ultimately the 1665 plague did not spread to Leicestershire that summer, but Palmer's collecting suggests that just the rumor
of an epidemic sparked discussions and recipe exchange.[36] The four attributed recipes show that the family not only looked among extended family

and friends, such as "Cousin Adderley" and Mrs. Whatton, but also turned to learned physicians such as Dr. "Micklethwayth."[37]

The Palmers and the Bennetts, of course, were not alone in gathering medical information in reaction to particular bouts of sickness within their own households. While recipes of this sort make up only a small portion of any given collection and account for a small area of a family's collecting practices, this type of reactive compilation is widely evident. Within the bundle of loose recipes in the Glyd/Brockman papers (discussed further in chapter 5) is one for an "opening drink that is very good to drink about a month before delivery." The recipe advises the maker to boil a range of plant materials (half a dozen roots, a couple of kinds of seeds, and a handful of various fruits) in water and wine to make a concentrated liquid that was then to be mixed with a little sweet almond oil. Dated 1686 and subsequently attributed to a Mrs. Streats, the recipe arrived in the Glyd/Brockman collection just before Anne Glyd Brockman's sister Martha gave birth to little Martha Drake.[38] Another letter within the collection, written by Anne Glyd Brockman's daughter Ann Boys, offered a recipe for "a water very good for ani heates." This came with a personal plea from Boys, who wrote, "pray mother let ther be sume of this water made and take it I am perswaded it will doe you a greate deale of good."[39] In these cases, sharing health-related information was a way family members expressed their concern and affection. Here compilers were collecting not only *with* their family but *for* their family. Given the size and scope of many surviving household recipe books, it is unlikely that their compilers used or witnessed every recipe in the book. In that sense recipe books cannot be taken as direct reflections of a household's experiences with sickness. However, the health concerns of a particular household undeniably shaped their collection of recipe knowledge in profound ways.

Calling on Family and Kin

Extended family and kin, variously described, also contributed to a recipe collection in other ways. Within the Palmer and Bennett collections, a significant portion of recipes are attributed to donors described as family members: sisters, aunts, cousins, sons, daughters, and in-laws. Examples include the recipe for treacle water given to Archdale on 25 May 1666 by "Sister Gore" and the two recipes, one for a surfett and another to ease the pain of a sore breast, by "Sister Mary Palmer."[40]

Practical know-how from family and kin entered early modern recipe notebooks by a variety of paths. As we saw earlier with the Glyd/
Brockman family letters, recipes circulated within the large family and
household correspondence networks. The collection of Elizabeth Freke
(1642–1714), for example, contains a letter from her sister Lady Austen
enclosing Lady Powell's recipe for laudanum.[41] Accompanying the recipe
is an account of Austen's own positive experience with the cure, including
precise dosage, and her recommendation that her sister try this medicine.

Aside from correspondence, face-to-face visits also were fruitful occasions for recipe exchange. Palmer's collection suggests that guests at Wanlip Manor were often called on to contribute to his ever-growing compendium of medical information. When his daughter-in-law Elizabeth
D'Anvers Palmer visited Wanlip in late June/early July 1662, she gave
Archdale six recipes for accomplishing household tasks such as making
mustard, roasting a pike, stilling snail water to treat consumption, and
washing ribbons.[42] When Elizabeth's father, William D'Anvers, visited
Wanlip a little over a year later, he also added to Archdale's collection
with a recipe for "liquoringe of Bootes."[43]

Within the *Arcana Fairfaxiana* there are several recipes from Mary and
Henry Fairfax's relatives. For example, Mary's brother Henry Cholmeley contributed a large number.[44] While many of these were copied out
in Henry Fairfax's neat hand, Cholmeley also wrote in recipes himself.
For example, he wrote out and signed his name under "A certaine remedy for the toothach if it proceede from heate."[45] Henry Fairfax's own
brother Ferdinando also wrote a recipe "For a could" into the book.[46] Unlike Henry Cholmeley, Ferdinando neglected to sign his name, and the
attribution was later provided by the diligent Henry Fairfax. Perhaps returning the favor, Fairfax himself contributed to Ferdinando and Rhoda's
book, where he wrote, in his distinctive hand, recipes for making "Walnutt
water" and a cordial water.[47] The recipes the brothers wrote in each other's books do not overlap with information in their own household books,
and it is difficult to reconstruct the exact circumstances that prompted
these entries. However, they emphasize the importance of family collaboration in recipe compilation and the openness of household books to contributions from family members outside the nuclear unit. For future generations, these two recipes documented the close relations between the
two brothers and their families.

In early modern England, kinship terms held a wide range of meanings
and were fairly flexible.[48] Parents might address their offspring's spouses

as sons and daughters, and siblings might call their siblings' spouses brothers and sisters. In this way terms such as "sister" could signify a variety of
relationships, from biological sister to sister-in-law to half-sister to step-
sister. "Cousin" was particularly widely used and could apply not only to
the offspring of aunts and uncles but also to more distant kin. The term
was also used more generally to establish that a kinship claim existed.[49] In
some cases, with additional biographical information to hand, we are able
to reconstruct the exact relationship between recipe donor and compiler.
For example, Elizabeth Freke took recipes from her siblings, aunts, and
cousins. These included recipes from "my cousin Penrudock," two recipes
from "my aunt Tregonell," and recipes from her sisters Lady Austen and
Lady Norton.[50]

In many other instances, however, we lack the relevant information
to discover these relationships and networks and thus have to extrapo-
late, from these brief descriptions, the significance of the citations. Here
Naomi Tadmor's detailed analysis of how kinship terms were used in
early modern England provides a good framework for reading our recipe titles and attributions. She argues that the broad meanings and wide
range of relationships associated with terms such as "cousin," "sister," and
"brother" constituted one way families achieved inclusivity and equality.
The language of kinship, she writes, "was employed habitually in a wide
range of interpersonal relationships to claim recognition, propose social
bonds, set moral and religious duties, and postulate many expectations."[51]
Thus, while further details on recipe source attributions are useful, it is
perhaps more helpful to think broadly about the citation of family members. The strength of kinship bonds may have brought certain assurances
about the trustworthiness and reliability of the recipe donors. The family
bond may also have allowed compilers and donors to exchange recipes
within a fairly limited network. That is, recipes, as valued knowledge,
could be safely communicated and still be kept "within the family." Given
the substantial role that kin played in the lives of many households, it is
not surprising that they collaborated on such a scale.[52]

Friends, Neighbors, and Recipes

Recipe collectors such as Palmer and the Bennetts not only called on
family and kin for recipes but also depended on their friends and neighbors. Looking outside the family, this section explores the myriad ways

our historical actors turned to their extended kin, friends, and acquaintances to acquire practical know-how. Many, as I will show, both cultivated local social relationships and used them to extend their recipe knowledge. Social structures, networks, and alliances formed one of the main frameworks for exchanging recipes.

Recipes, Reading, and Sociability

Social gatherings were fertile ground for gathering recipes, and compilers' existing social and local networks crucially shaped their access to such knowledge. By all accounts, Palmer seemed to always be on the lookout for new recipes and had a knack for turning dinner conversations toward household know-how.[53] On 13 May 1663, after the Northamptonshire minister Mr. Hollid dined with the Palmers, he returned the kindness by leaving Archdale with two recipes: one "For a bone of pinne in the throate" and another "for the Ricketts."[54] When Palmer himself visited friends and family, often dining with them, he continued to press for recipes. For example, he collected instructions for a remedy for a fever from "Cousin Peacocke" after a dinner at "Father Smith's" on 12 March 1662/63.[55] When he visited Dr. Bowles and his wife in Oundle on 16 April 1659, he returned home with recipes for cordials and two remedies for sore teeth and mouth cankers in children.[56] A visit to his old university town, Cambridge, provided instructions for growing asparagus, confided by William Moses, master of Pembroke Hall.[57]

Social acquaintances could also extend one's network of recipe knowers and pass on know-how they themselves had collected from others. For example, one recipe in Palmer's collection is titled "2[d] Receit for a Burne or a scald. Given mee by one Eliz: Janes 8th Janu: 1669/70 Kinn to Eliz: Woodcock Onelepp."[58] It was not uncommon to receive recipes at second or even third hand. On 9 March 1663/64, for example, Palmer received a recipe for scurvy from his cousin Peacocke labeled "Sir Geo: Penrudducks Receite given of him by Lo: Ruthen & given mee by Cous: Peacocke 9th March 1663/4."[59] The connection of multiple authors with single recipes represents both the compilers' interest in the many hands a recipe had experienced and their recognition of collaborative knowledge making. The notion of authorship in relation to these recipes is more complex than individual creatorship or proprietorship. Aside from personally offering recipe knowledge, family members could also host gatherings where recipes were exchanged. An entry in Palmer's book is described as

"For any Ague but onely a Quartan. Given to mee by Cous: Gabriel Taylor. 8th June ano: 72 in Sister Stanleys house in Leicester."[60]

Informal social and knowledge networks also brought access to unfamiliar texts. Large-scale copying of existing manuscripts, when accessible, was fairly common. One interesting illustration is the recipe collection of James Tyrrell (1642–1718).[61] Tyrrell, a political theorist and historian and a friend of the natural philosopher John Locke (1632–1704), composed a small notebook of recipes with extracts taken from several sources. The book begins with the following inscription: "These following Receipts I had among Mr Hartlibs collections in the possession of my Lord Brereton in his Library at Brereton Hall 1685"; this is followed by a section headed "Receipts from Mr Lock's Collections 1691."[62] The inscriptions not only betray Tyrrell's long-standing interest in recipe knowledge but also provide a rare glimpse into his personal consultations of the Anglo-Prussian intelligencer and reformer Samuel Hartlib's (c. 1600–1662) papers, which included his *Ephemerides* and extensive correspondence with virtuosi such as Robert Boyle, William Petty, John Aubrey, John Beale, and others.[63] The papers and their afterlives have a complicated history—by 1667 they were in the library of William Brereton, third Baron Brereton of Leighlin (1631–1681) at Brereton Hall in Cheshire, where Tyrrell encountered them in the 1680s.[64] Tyrrell's small selection from Hartlib's papers included a number of recipes taken out of various letters, including, for example, a recipe for "Spiritus Calcis Vivae" by Basil Valentine, sent first from Sir Kenelm Digby to Robert Boyle and then forwarded to Samuel Hartlib and his son-in-law Frederick Clodius (1629–1702) for production. Another example is the set of recipes sent to Hartlib by Boyle's sister, Katherine Jones, Lady Ranelagh, that include Sir Kenelm Digby's "secrets" against "festers and inflamations" that came with "extraordinary commendation."[65] Like many other compilers, Tyrrell had particular medical interests—in his case it was cures for kidney stones. Here Tyrrell copies a run of recipes to ease suffering from kidney stones gathered from a number of sources, from Dr. Mackellow to "Mr Wood," perhaps the antiquary Anthony à Wood, to a London merchant.[66] Nine years later Tyrrell expanded his collection of recipes with a lengthy selection out of John Locke's collection. This section again consisted of recipes by a wide range of contributors, from unknown women such as Mrs. Clerk, Mrs. Higgins of Bristol, and Mrs. Jep to well-known authors such as Thomas Sydenham and Robert Boyle.[67] Tyrrell evidently began to see his reading notes from these two collections as a unit, constructing a partial table of contents spanning both these extracts.[68] Tyrrell's transcription of the Hartlib Papers occupied a mere twelve folios, emphasizing the strict selection that occurred during copying. Even when copying from a reputable source, compilers chose what they deemed most useful and suitable for their own needs.

Recipe gatherers not only were keen to get their hands on manuscript texts but also relished and took advantage of temporary access, often offered during social or family visits, to expensive or harder-to-obtain printed books. When Elizabeth Freke stayed with her sister Lady Frances Norton and her niece Lady Grace Gethin,[69] she made extensive copies from two contemporary pharmacopoeias—William Salmon's translation of George Bate's *Pharmacopoeia Bateana*[70] and the English translation of Moise Charas's *Pharmacopée royale galenique et chimique*, which was first printed for John Starkey and Moses Pitt in 1678 as *The Royal Pharmacopoea, Galenical and Chymical*.[71] Within Freke's notes from these two volumes, recipes from the printed texts are intermingled with entries on the same ailments from her sister Lady Norton and her niece Lady Gethin. The clustering of these recipes suggests that Freke paired her reading of the two printed volumes with family conversations on similar topics. The results of this reading and informal discussion among the three women can be seen in the alternating arrangement of information gleaned from print and oral sources. Freke's reading of printed medical books is thus embedded within her social milieu. Here printed books not only offered new information but also sparked dialogues within her household.

Local Networks of Recipe Knowledge

Whereas Palmer's notes reveal his wide-ranging social engagements, the names of recipe donors in the Bennett notebook paint the picture of a family with strong local links. In 1668 Philip Bennett the elder purchased South Brewham Manor from Francis Swanton, and by the second generation the family appears to have been active in the local community. Philip Bennett the elder and younger both served as clerks of peace of Somerset. Philip the younger was also treasurer of the Wincanton Fire Relief Fund in 1707.[72] Scanning the place-names cited in the notebook shows that a large proportion of the recipes were collected from individuals living in towns and villages surrounding the Bennetts' home in Brewham. More than ten recipes in the collection were gathered from friends and relatives living in their hometown of Brewham, the neighboring towns of Bruton and Batecombe, and the village of Kilmersdon. Almost thirty towns and villages are listed in the Bennett family recipe collection, most of them within the corner of the country where modern-day Somerset borders Wiltshire. The majority of the places cited, such as Bruton, Lamyatt, Upton Noble, Castle Cary, Stourton, and Wincanton, are within ten miles of the Bennetts' home, suggesting that they tended to gather recipe knowledge from members of their local community.

Two readings of "proximity"—geographic and family—framed these recipe-collecting practices. In the Bennetts' case, since Bruton and Batecombe were both less than five miles from Brewham, proximity of location accounts for how frequently they appear in the notebook. Kilmersdon, however, was approximately fifteen miles away and was one of the farthest villages from the Bennetts' base. In this case it is likely that family relationships account for the town's repeated presence in the notebook, since the entries are attributed to "Cousin Margaret Trowbridge" and "Sister Shute." Social visits with relatives at Kilmersdon not only allowed the Bennetts to tap into the medical knowledge of their hosts but also introduced them to local experts and knowledge networks. As described above, Katherine Bennett's sickness in Kilmersdon in 1713 brought the Bennetts into contact with Dr. Rowland Cotton. The collection also contains information gleaned from other sources in Kilmersdon, including Thomas Maggs and Jeane Clarke, suggesting that the Bennetts found visits to the village fruitful for recipe gathering.[73] Aside from members of their extended family, a large portion of the "local" recipes cited in the Bennetts' notebook were gleaned from friends, acquaintances, and even passersby. In some cases, such as a recipe taken from "a man of Froom" or "A woman at Major Horners House at Mells," it appears that the Bennetts did not even recall the contributors' names. The lack of attribution here could be due to factors from a lack of familiarity to the contributor's low social status.

However, geography did not totally limit householders' access to medical information. Palmer, for example, traveled extensively around his county and ventured farther afield into the country, gathering recipes and using social visits to tap into "new" networks of knowledge. When Palmer visited his kinswoman Mrs. Palmer in Aston-upon-Trent in Derbyshire on 28 July 1659, he received from her a remedy for toothache that she herself had obtained from Dr. Wright of London.[74] The regular travel patterns of the wealthy and landed social groups (to which many recipe book creators belonged) meant that compilers could be part of several "local" networks. A case in point is Johanna St. John, familiar to us from the introduction.

As we will learn in more depth in chapter 2, St. John maintained a steady correspondence with her steward Thomas Hardyman during the 1650s and 1660s. Largely based in London, the St. John family relied on their country estate in Lydiard Tregoze for provisions, including ingredients (such as butter) for medicines. The letters reveal that St. John directed Hardyman to ask his wife and the local apothecary, Mr. Goram, to distill waters, to apply to Goody Wolford for herbs, and to plead with Bess the dairymaid to produce enough butter for St. John to make her balsam. Clearly, despite being primarily based in London, St. John was embedded in the local informal knowledge networks of Lydiard. Yet while St. John appears to have relied on her communities of medical knowers in Lydiard Tregoze to produce medicines and foods, these figures do not have a strong presence in her surviving manuscript recipe notebook. Instead, the book is closely tied to London court circles and fashionable physicians. Robert Boyle (1627–1691), his sister Katherine Jones, Lady Ranelagh (1615–1691), and the physicians Thomas Willis (1621–1675), George Bate (1608–1668), and Richard Lower (1631–1691) were all frequent contributors to the volume.[75]

Additionally, St. John collected recipes from members of her Battersea household, including Mrs. Patrick, wife of the family's resident chaplain.[76] St. John was thus participating in two local knowledge networks. While living in London, she was actively exchanging recipes with notable physicians and acquaintances, and through an intermediary she maintained strong contacts with her other local network at her country estate. The absence of the humbler Lydiard-based knowers in St. John's recipe book suggests a certain amount of editing on her part, perhaps even a self-conscious fashioning of the recipe book into an object reflecting her elevated social network in London. What is evident here is that the knowledge networks supplying compilers with medical information depended greatly on the social contacts, travel habits, and life patterns of individual compilers. The importance played by local contacts reinforces arguments put forward by other scholars on the importance of early modern communities and neighborliness.[77]

Local knowledge thus seems to have played a crucial part in the assembling of the early modern recipe books. Yet knowledge gathered locally does not necessarily translate to definable "regional" medical practices or knowledge. Anne Stobart, in her study of domestic medicine in the south-west of England, concludes that, despite the geographical parameters of her study, it did not paint a picture of regional medicine.[78] Stobart's in-depth study of the private papers, household accounts, and medical recipe collections of four late seventeenth-century families based in the south-west of England suggests that the resources, social connections, and at times medical needs of many families took them on journeys to London or other major cities. The sophisticated routes of knowledge transfer via vernacular print, correspondence, and travel meant that recipe knowledge circulated widely. The scope of "venturing out" of their local knowledge networks undoubtedly varies from one compiler to the next. Many com-pilers, much like Archdale Palmer, continued to collect recipes while away from home, bringing in knowledge and ideas gleaned from beyond their immediate geographic or social vicinity. That is not to say there are not distinctive characteristics that differentiate the medical landscape of dif-ferent regions or subregions, as Steven King and Alan Weaver have ably shown for eighteenth-century Lancashire, but rather to note that such lo-calism is hard to detect in the family recipe books.[79]

Recipes and the Gift Economy

The central role played by existing social, kinship, and particularly local networks encourages us to read recipe transfer alongside exchanges in other economies and as part of a larger framework of obligation and patronage.[80] Recently a number of studies have highlighted the impor-tant role played by medical gifts. In particular, Alisha Rankin's study of pharmacy in the sixteenth-century German court argues for the need to view the exchange of recipes, materia medica, distillation equipment, and the like within the larger "economy of scientific exchange."[81] Similarly, Meredith Ray has written of the flurry of recipes, perfumes, and more exchanged among women in Renaissance courts and in particular in the circle of Caterina Sforza.[82] In the early modern English context, gifts of recipes likely functioned within what economic historians have termed "economies of obligation," where social and moral credit or reputation worked alongside economic credit to delineate financial, material, and knowledge exchanges.[83] Household members consequently participated in complex and overlapping networks of reciprocity and obligation. The correspondence between Edward Conway (1594–1655) and Edward Har-ley (1689–1741) furnishes a clear articulation of this practice. In 1651 Con-way wrote in a letter to his nephew Harley:

> My Lady of Westmerland hath two Receipts the one the preparation of Oxe
> Galles, the other A water of Mirrhe, both of them washes for the face, if you
> have so mutch credite with hir I pray get them of hir, and communicate them to
> mee they are very excellent and you will thanke me for them.[84]

In this instance it turns out that Harley's credit with Mary Fane (c. 1611–
1669), wife of Mildmay Fane, second Earl of Westmorland (1602–1666),
was sufficient, since a few months later he sent the requested recipes.[85]
The explicit mention of credit here reminds us of the high value placed
on both social reputation and practical knowledge. Access to a family's
recipes was not granted to just anyone, but only to those who had proved
worthy. In fact, granting access to one's own practical knowledge was akin
to a gift, a favor that must be recorded and returned. Certainly, within
the Conway/Harley correspondence, a recipe was offered with the expectation of repayment in kind. For example, in a letter dated 14 January 1650/51, Conway acknowledged Harley's gift of recipes by writing, "I
thanke you very mutch but will thanke you better for I will send you some
in exchange." In a later letter Conway wrote specifically of being "in debt"
to Harley for both his letter and his recipes.[86]

    Offers of single recipes were also made in a bid for patronage or other
goods. Within the Bagot papers at the Folger Shakespeare Library, for example, is a letter by Mercurius Patten offering two recipes for deafness
to his patron Sir Walter Bagot in hopes that Mrs. Bagot might bestow a
piece of bacon on his household.[87] In these cases the value of the recipe
is made clear as recipe knowledge is exchanged for everyday goods. Contemporary customs and patterns governing gift exchange and economies
of credit and obligation were therefore central to the transfer of recipe
knowledge.[88]

    Besides gifts of single recipes, it was also common to present entire
collections of recipes as gifts. The producers of printed recipe collections
adopted the common early modern practice of using dedicatory epistles
to garner favors and seek patronage.[89] For example, the popular *A Choice
Manual of Rare and Select Secrets in Physick and Chyrurgery* is dedicated
to the "vertuous and most Noble lady" Letitia Popham, wife of Colonel
Alexander Popham. Within the letter the book producer, W. J., offers this
"small Manuall, which was once esteemed as a rich Cabinet of Knowledge," as a gift worthy of Mrs. Popham's patronage. The gift, according to
the dedicator, was his way of expressing gratitude for past favors granted
by Colonel Popham.[90]

While book producers might have relied on book dedications to position recipe collections as gifts, early modern householders could give actual notebooks as presents. Many examples can be found in the recipe archive. In one case "Poore Colly" offered her patroness a presentation copy of a treasury of recipes by another woman, Hester Gullyford. Colly's inscription reads, "This book humbly beggs Madam Elizabeth Butler her acceptance from her faythfull servant, March the last 1679, Poore Colly."[91] Another example is the recipe collection of Cecilia Haynes Mildmay, which has "Ci Mildemay 1663" written on the first folio and "Lady Anne Lovelace gave me this book Cisilea Haynes 1659" on the recto of the second folio.[92] Cecilia evidently so treasured Lady Anne's gift that she reinscribed it and carried it to her new home upon marriage. In both of these cases we might locate the gift of a recipe collection alongside other presents commonly exchanged by English aristocrats and gentlewomen. Certainly a beautiful leather-bound, embossed recipe book is not only a gift of treasured recipe knowledge but also a representation of the resources required and care taken to produce such an object. As such, within the gift economy such books probably functioned like other lovingly handmade objects such as marmalades, embroidery, and made-up medicines often used to consolidate female alliances.[93]

Significantly, many recipe collections circulated far beyond immediate family members and were pored over by neighbors and acquaintances. For example, lending one's collection to others, as Florence Mompesson shared hers with the Bennetts, gave the borrower of the book a chance to closely study one's collection. As we saw above with the Fairfax family, neighbors, friends, and relatives often copied entries from a family's compendium of recipes, and visitors were encouraged to write their own offerings straight into the notebook. For these inscribers and readers, the book at hand must have offered both medical knowledge and crucial social information. The numerous names cited with individual recipes would have acted as a display of the owners' own networks and the far-reaching (they hoped) extension of their social credit. By inviting these "outsiders" to share their recipe books, families were also inviting them to witness, marvel at, and appreciate their hard-earned social alliances and family networks.

Recipe collections, then, were afforded a dual function within early modern households. On one hand, they were repositories of household knowledge. On the other hand, they were maps of a family's social network.[94] Once circulated among friends and family, the recipe book also served as a public record of a family's treasury of health and social alliances. To those scanning the various recipes alongside their donors' names, these books at once proclaimed the wealth of compilers' credit networks and visibly demonstrated their standing within the community. If we look at this from the recipe donor's perspective, to have one's recipe included in the same collection as that of an illustrious aristocrat or a morally upstanding resident of the local village was a way of joining a particular club. Within this framework, existing notions of social credit and moral reputation shaped recipe transactions. Notebooks of recipes, culinary and medical, were thus records of these transactions—a ledger of credit and debt and a tangible demonstration of a family's overlapping networks.

Venturing Out: Seeking Knowledge from Experts

In their search for new recipe knowledge, householders often ventured outside their family and social circles. In the Bennett family book, for example, a dozen recipes are attributed to individuals described as "goodman," "goody," or "goodwife." These included a recipe for the ague from Goodman Gerhead Wedmoore on 11 June 1716 and a recipe for the fits in a child from Goody King of Sturton given on 5 May 1707.[95] The Bennetts were not alone in recording recipes from this particular group within the community; other gentry compilers such as the Carey family also considered the knowledge of local men and women worthy of putting on paper.[96] During the period, these terms were used selectively to refer to individuals below gentle status in the social stratum who made significant contributions to the affairs of the parish. David Postles argues that they "had an almost honorific import, and demarcated a select status within a social level, based on service and neighbourliness."[97] Thus, while the medical knowledge of apothecaries was worth writing down because of their occupation, the knowledge of Goody King, for example, was considered useful because of her position within the community and her general neighborliness. Perhaps thinking along the same lines, Archdale Palmer collected recipes from several clergymen including Mr. Jenner, described as a minister in Ireland, Mr. Hollid, a minister in Northamptonshire, Mr. Smith, the "late Minist. Of Oneleppe," Mr. Campion, a minister in Shropshire, and finally Mr. Sedon.[98] In this sense the compilers' assessment of the "knowers" of practical information is, as Craig Muldrew has demonstrated for credit and obligation, based on social reputation and moral standing.[99]

When it comes to medical recipes, we commonly find entries attributed to persons labeled "Dr" in the recipe books. "Doctor" was a somewhat ambiguous descriptor in the early modern period. Many contemporaries employed it to denote any person practicing medicine rather than only university-trained physicians, as well as those trained in theology.[100] However, within the recipe archive there are also a significant number of instances where we can identify the cited medical practitioner, providing rich case studies of medical practice and, in particular, how householders combined home-based health care with services obtained on the medical marketplace.[101]

Prescriptions from physicians and other practitioners could have ended up in household recipe collections through several paths. Certainly some were provided in face-to-face consultations. For example, as I noted earlier, Johanna St. John's recipe book contains entries from the well-known physicians and natural philosophers Thomas Willis and Richard Lower. A number of these are clearly the result of medical consultations.[102] For example, one recipe is labeled "A Snayl water Dr Willis prescribed me when I was with child for sharpnes in my blood." This was followed by a series of remedies linked to pregnancy and childbirth, many attributed to Willis, suggesting these might be also from direct consultations.[103] Likewise, St. John seemed to have been a patient of Dr. Lower. In the margin of a recipe titled "Sr Edward Greaves For the Scirvies" she wrote, "Dr Lower bid me add powder of perl & corral & crabs eyes to this of each half an ounce."[104] St. John also recorded a recipe titled "Syrope of Steel Dr Lower gave me & which he uses."[105]

Another example can be found in the recipe book associated with the family of Henry Carey, second Earl of Monmouth (1596–1661), and his wife, Martha, which contains detailed instructions given by Dr. George Bate (1608–1668) for Henry and Martha Carey and by Dr. Theodore Mayerne (1573–1655) for their daughters Theophila and Mary in the mid-1650s.[106] Both Bate and Mayerne were active members of the College of Physicians and fashionable physicians at the time. Whereas Mayerne was connected to the Stuart court (he was physician to both James I and Charles I), Bate was the physician to Charles I and Charles II and also to Oliver Cromwell.[107] The recipes in question are titled "Dr Mayernes Method for Lady Theophila Cary's Cold 1654 May" and "Dr Mayerne's method for the Lady Mary Cary's cold May the 17th 1653."[108] Both these entries recommended a series of medicines and diet drinks for the patients as well as giving precise instructions for phlebotomy.[109] The entry for Lady Mary Carey's cold, for example, included a "pectorall decoction," a purge, a recipe to make a china broth, and a recipe for lozenges for the cough, which Lady Mary was instructed to melt under her tongue.[110]

Since patients often corresponded with their medical practitioners, it is also not unusual to find original consultation letters preserved alongside recipes.[111] Anne Meyricke, for example, copied out instructions addressed to her from her physician Dr. John Powell. In his instructions, Powell advised her to obtain from the apothecary some "antiscorbutick electuary" and to take it every morning with a large quantity of nutmeg. In addition, he told her to drink a posset made with white wine and scurvy grass, brooklime, and horseradish root. Finally, he recommended she take "Bath" water and gave advice on which apothecaries should make up her medicines.[112] Another example of this type of correspondence is the letter William Pearce, a surgeon, sent to Sir George Oxinden on 24 March 1662, where he included remedies for gout. A few weeks later Pearce followed up this letter with an inquiry about his patron's health, with further instructions, and added a postscript that his "Brother Bale is very well versed in those distempers and have good receipts for the same if occasion require."[113] While the examples cited here tend to stress householders' reception of recipe information, we should also bear in mind that physicians and other medical practitioners likely participated in wider, perhaps socially based, exchanges of recipe knowledge. For example, Michael Stolberg has recently shown us how the learned Bohemian physician Georg Handsch (1529–1578?) collected much health-related information from his own family members and their patients.[114]

Many of the recipes originating from personal medical consultations were passed around to friends and family. In one instance the Bennett family received a recipe for a scurvy remedy from Mr. Sam Boord of Batecombe on 22 April 1710. Mr. Boord was himself taking this remedy on the advice of "his Doctor Warmister."[115] Undoubtedly, when Dr. Warmister offered Sam Boord this particular prescription, it was within the regimen he specifically designed for the unique humoral composition of Boord's body at that moment. For Boord and the Bennetts though, the cure was no longer tied to a particular body on a specific occasion but rather was universalized and ready to be tried and applied to others with similar symptoms. In other words, there was a move from focusing on individual human bodies to addressing particular ailments and diseases across the population. Thus prescriptions that physicians and other medical practitioners gave to their patients on specific occasions entered the general realm of recipe exchange. Taken out of the context of the particular regimen they were created for, these recipes became common knowledge and were applied to patients with similar complaints.

Early modern compilers' desire to gather and record information from doctors needs to be read alongside the know-how they gathered from other experts.[116] It was common for compilers to seek information from those whose occupations gave them practical hands-on experience in certain areas. For example, Palmer's notebook includes a recipe for tobacco obtained from "Mr Clark an Apothecary at Lugborow" while he was staying with his sister D'Anvers at Swithland,[117] a recipe "for a bone spaven" at "Templehall" from a farrier,[118] and two recipes for making stone bottles sweet from "Fran Pochyn of the Mitre Taverne in Cheapside."[119] The Bennetts' book has recipes from Hannah Draper, described as "the Doctors old servant."[120]

In these particular knowledge transactions, we see the recipe compilers searching for useful information outside their own social and family circles. The need for the "best" set of instructions for particular household tasks encouraged compilers to seek out the widest possible pool of knowledge providers. Palmer's and the Bennetts' records of these donors' occupations suggest that both compilers viewed occupation as central to assessing the donor's expertise and claim to authority. For Palmer, an apothecary, a farrier, and a tavern keeper were certainly experienced in making medicines, caring for horses and cattle, and keeping bottles smelling sweet. Undoubtedly, in the eyes of the Bennetts, Hannah Draper's occupation as a "Dr's" servant afforded her opportunities to gain specialized medical knowledge. Including knowledge from these donors emphasizes the value recipe collectors placed on experiential knowledge.[121] The importance of hands-on experience is a theme we will encounter again and again in this book, particularly in later chapters focused on the multistep codifying of recipe knowledge and on trying and testing. Here I note that the high worth accorded to hands-on experience with particular cures or practices prompted compilers to cross social and gender lines in their recipe collecting. Furthermore, this venturing outside one's social circle also follows arguments put forward by social historians such as Keith Wrightson and Phil Withington.[122] The latter convincingly argues that there were particular spaces, such as taverns, where men and women of different social sorts mingled. Within this context, it is not surprising that Archdale Palmer came away from the Mitre Taverne armed with recipes.

Conclusion

Through detailed readings of the Palmer and Bennett recipe books, this chapter has offered a sustained look into the recipe knowledge and social worlds of two early modern families.

The compilation practices of these two families highlight the early modern fascination with and need for practical information such as recipes and the informal methods of gaining this knowledge. With their chronological organization, both of the notebooks at the center of this chapter resemble personal journals more than contemporary medical formularies or pharmacopoeias. By paying close attention to the records of recipe exchange, this chapter has explored how such compilers went about filling their notebooks with recipes and how they recorded recipe knowledge. For the Palmer and Bennett families, collecting recipes was part of daily life, and sociability occupied a central place in its shaping and framing.

In Archdale Palmer's notebook, dining rooms, stables, and public taverns all feature as sites for gathering recipes. Consultations with medical practitioners aside, family visits and social calls on friends and neighbors were occasions where medical information and advice were discussed and exchanged. Social visits also brought opportunities to examine and explore new libraries and transcribe passages from manuscript and printed texts into one's own recipe book. For the keen recipe gatherer, it also opened doors to new social and knowledge networks. Access to trusted and specialized information, whether from texts or from experts of good repute, was a crucial factor in transferring recipe knowledge. Sociability played a vital part in shaping this access.

Given that requesting and sharing such knowledge was conducted within the web of social and family relationships, we might consider recipe exchange as an act that created bonds between seeker and donor. After all, to apply for advice, recipe seekers must also confess their ailments and bodily dysfunctions. As recipe donors accept seekers into their knowledge networks by sharing their intimate health issues, recipe seekers also welcome donors into their circle of knowers. The actual processes of collation—asking for recipes, writing them down, and sharing them— are thus in fact social and cultural acts that tie recipe seekers and recipe donors together in new communities and consolidate existing networks and relationships. Consequently, for early modern householders, recipe knowledge represented more than just useful advice on how to alleviate sickness and promote health or to bake the perfect cheesecake; rather, it served a multitude of social and cultural roles. Recipe books, then, should be read as medical, social, and cultural artifacts.

Throughout the chapter I have also emphasized that creating recipe knowledge was collective, collaborative, and often cumulative work. This had a number of consequences. First, as we saw in the case studies above, a family's collective work on its book of recipes often spanned generations. This encourages us to recognize that the texts are largely multiauthor and difficult to date precisely.[123] Second, it becomes clear that household recipe knowledge is shaped and framed by family and marriage alliances in addition to connections such as institutional frameworks or social and knowledge networks. Not only did these family ties influence knowledge sources in the form of recipe contributors or recipe exchange networks, but, as we will see in chapter 4, they also had epistemic impact, particularly on trial, testing, and recipe evaluation. Finally, the active involvement of early modern men such as Archdale Palmer, Henry Fairfax, and others reminds us that recipe knowledge interested men and women both. Although historians have often connected the collection and use of household knowledge with the domestic sphere, housewifery, and the female domain, it appears that such knowledge and practice cannot be so neatly categorized. The continual crisscrossing over gender lines in the transfer of knowledge suggests that the gendering of such knowledge was nuanced and complex. The gendering of housework, of which collecting recipe knowledge is firmly a part, was often influenced by the idiosyncrasies of individual partnerships or was in flux, responding to changes wrought by different life stages or economic circumstances. As we will see in the next chapter, in many cases the complex task of ensuring the smooth running of an early modern household required a team of men and women armed with a range of skills and expertise.

CHAPTER TWO

Managing Health and Household from Afar

King Charles II wanted Muscovy ducks. It was the spring of 1662 and, inspired by his French sojourn, the king was in the midst of redesigning St. James's Park. With the new canal in place, he now sought Muscovy ducks, preferably white ones, to complete the water feature. Undoubtedly his wishes were whispered throughout the court, and courtiers and landed gentry of the day scrambled to furnish the waterfowl he wanted. Lady Johanna St. John, whom we met in the introduction, and her husband, Sir Walter St. John, were one of couples in this race. In two letters, both dated 13 March 1662, Johanna pleaded with Thomas Hardyman, steward of the St. Johns' country estate, Lydiard Park, to "buy [or] if you cannot beg" for at least a pair of these ducks. After providing Hardyman with a couple of local leads, Johanna particularly instructed him not to say "for who[m] we would have them for then if they be people of any qualitye they wil present them themselves."[1] Luckily for the fortunes of the St. John family, Hardyman came through, although one of the ducks died in transit from Wiltshire to London. Johanna's final letter on the subject begged Hardyman to "present Sr W['s] service and mine to the persons of whom yu had them" and to "send no more unles you can git any whit ones for the K desirs much some white."[2]

By 1662, the time of the Muscovy duck "crisis," the St. Johns had settled into a routine of spending time in London while Parliament was in session and, as many families of the gentry did, going back to the country for the rest of the year. The St. John family headed to their vast estate in Lydiard Tregoze near Swindon. Just a few years earlier in 1656, at age thirty-four, Sir Walter St. John had inherited his father's title and estates, including the properties in Battersea and Lydiard. In the same year Sir Walter also became the member of Parliament for Wiltshire for the second Protectorate Parliament. These two events shaped the way the couple lived for the rest of their lives. For the next few decades Sir Walter continued to pursue a career in politics in London, and the family primarily resided in Battersea. The vast estate of Lydiard Park became both a country retreat and a working estate supplying the London household with provisions. Muscovy ducks were not the only things that traveled between Lydiard Park and St. John Manor in London. The road between the establishments was well worn with letters, foods, medicines, household items, and the St. John children trekking back and forth all year round.

The two letters concerning the St. Johns' desperate search for rare waterfowl to garner favor with the king are part of a series of about eighty letters between Johanna and Hardyman concerning the day-to-day running of the St. John estates. Packed with unusual detail about mundane concerns and anxieties, these letters reveal the complex work of managing an elite household in mid-seventeenth-century England. Drawing on the St. John letters, this chapter explores the intersection between household management and planning, household science, and the care of a family's socioeconomic and political "health." In doing so, it reconstructs the rich context shaping the production and use of early modern household recipe knowledge.

Lydiard Park and Household Management

As a turn-of-the-century drawing shows, the St. Johns' country estate was a large country house with extensive formal and kitchen gardens and a swan-filled lake (fig. 2.1). Like many similar estates, Lydiard Park underwent a series of "improvements" in the mid-eighteenth century.[3] Johanna's detailed letters to Hardyman reveal that in the 1660s the main house and subsidiary buildings of the manor contained cellars, kitchens, a scullery, a larder, a still house, a dairy, and a pump chamber to store sugar and spices. The letters also mention a deer park, a rabbit warren, an orchard, a bowling green, and cultivated decorative gardens.[4] By all accounts, early modern country houses such as Lydiard were bustling places.[5] Inside the house, servants busied themselves cleaning, preparing meals, making cheese, butter, and other dairy products, and—depending on the season—preserving and conserving a range of foods. In the outhouses, teams of workers brewed

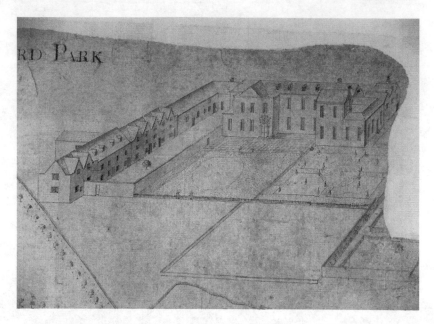

FIGURE 2.1. Plan of Lydiard Park c. 1700. Warwickshire County Record Office, CR162/714.

a variety of beers and other drinks, distilled and compounded medicines, kept turkeys, chickens, and other poultry, looked after sheep, oxen, and pigs, and cultivated vegetables, herbs, and fruits in the kitchen gardens and orchards.[6] As numerous studies have detailed, these duties were largely considered the work of female householders and servants.[7]

Analysis of prescriptive literature, letters, and accounts makes it clear that managing the household was also firmly seen, by both men and women, as mainly up to the wife.[8] Gentlewomen like Johanna St. John, aided by a team of upper servants such as the housekeeper and the steward, took on most of the responsibility of organizing and managing the operation. The tasks were wide-ranging. They included ensuring that the right sort of ale flowed freely on the St. Johns' tables; cultivating French and Portuguese melons; distilling various medicinal waters; and, as we saw earlier, sourcing white Muscovy ducks for the king to ornament St. James's Park. That is not to say the masters of the house were not involved in the day-to-day running of their households. Certainly in some families husband and wife shared the work. How far each contributed to the smooth running of the household largely depended on their inclinations and their marriage.

The letters from Sir Walter and Johanna to Hardyman reveal that for the most part they split the household management rather conventionally. Johanna, in her thirties by that time, took on organizing the provisioning, food preparation, and physic, while Sir Walter handled the household finances and keeping the male upper servants in line. In this the St. Johns seem to have reflected a division of labor fairly common among their contemporaries and often espoused by advice or conduct manuals.[9] Within the St. John letters, Johanna's voice emerges as the dominant one in household matters. Of the hundred or so letters the couple addressed to Hardyman, only sixteen are from Sir Walter. This may be chance survival or due to their mundane subject matter. Sir Walter wrote his share of letters to Hardyman, but they never reach the same specificity and concern about planning for provisions that Johanna's do. The letters are generally briefer than Johanna's, and Sir Walter tends to address Hardyman familiarly, as Tom or Thomas. They deal with a number of recurring topics: estate finances, tenants' welfare, and other estate matters; the purchase of horses; and staff management, with most detailing money owed and received from tenants and dependents.[10] A handful demonstrate that, like many of his peers, Sir Walter was both a caring and a strict master. In particular, the letter series contains two rather strongly worded letters berating Hardyman and other Lydiard servants for disobedience.[11]

With most written by Johanna, the St. John letters add a much-needed female voice to histories of household and estate management, shedding light on the managerial role taken on by elite women and emphasizing the range of knowledge, skills, and expertise the household collective needed to carry out these tasks.[12] Johanna's lengthy discussions of staffing dotted throughout the letters offer richly textured readings of contemporary notions of obligation and responsibility, and her detailed instructions to a group of dedicated experts, both male and female, offer complex readings of housework as gendered labor.

Three case studies form the core of this chapter. "Bess's Turkeys" examines a series of letters between Johanna and Hardyman concerning the rearing and fattening at Lydiard of turkeys for the London table, delving into contemporary practices about rearing poultry and training and managing servants. "Rudler's Melons" follows a series of conversations about the ornamental and kitchen gardens at Lydiard. By exploring the relationship between Johanna and her wayward gardener Rudler, I showcase the tapestry of skills held by various servants within the St. John household and tease out notions of obligation and the role specialized expertise and knowledge played in framing master-servant relations. Continuing the investigation of the intricate relations between employer and employee, the final case study focuses on distilling citron water and on the complex practices of home-based medicine.

A number of main themes weave through and connect the various cases. First, everyday domestic labor gave householders opportunities to investigate natural and material processes. Household tasks such as rearing livestock, cultivating plants, or distilling medicines relied on in-depth knowledge about the natural world, and by working through these processes men and women actively engaged with natural knowledge within their own homes. My reading of the St. John letters opens up a world where plants, animals, and other foodstuffs from around the early modern world were tasted, explored, tried, and experimented with daily.

Second, much as in the writing of recipe knowledge, household collectives equipped with a diverse set of expertise took on the tasks of household provisioning and producing food and medicine. These collectives included literate and elite gentlewomen and gentlemen like Johanna and Sir Walter St. John, who are highly visible in the historical record. They also included figures like Bess the dairymaid and Rudler the gardener, who are more reticent about their lives. This chapter brings to the fore these historical actors, often labeled "invisible technicians," highlighting their continued engagement with making natural knowledge and their multifaceted skill sets, which were used to ensure that foods and medicines were produced in the right quantity and, importantly, quality.

Finally, as the cases below demonstrate clearly, household collectives functioned within the traditional social hierarchies and notions of moral obligation; but, like recipes, these too were continually tested. As the case of Charles II's Muscovy ducks indicates, household management and food provision played a central role in contemporary sociopolitical systems. Foods, medicines, and know-how played central roles in contemporary systems of patronage and gift economies. As we sift through the quotidian minutiae of Johanna's letters, it quickly becomes evident that the provisioning needs of the family were shaped by the St. Johns' political and social ambitions. Turkeys, cheeses, and medicines not only were consumed by the St. Johns but also were shared with family, friends, patrons, and tenants. In this way Johanna's interests in the "health" of the St. John household extended beyond bodily health to encompass the household's financial position and social standing within the local communities and larger national networks.

Together these case studies attest to the necessity and importance of provisioning in large households, bringing to light the wide variety of tasks undertaken in country estates to ensure the social, bodily, and financial well-being of the landed gentry and the expertise and knowledge required to complete these tasks. This focus on self-production of food brings narratives on quotidian domestic labor and household economies into the conversation with existing studies on household consumption, which have tended to focus on acquiring goods and everyday consumables from shops and markets, largely in an urban setting.[13] As Johanna's repeated pleas to Hardyman reveal, quality control was paramount, and ensuring the "health" of the household required a delicate balance between in-house production and external sourcing such as cheese from a neighbor with particular gifts in dairy work and medical ingredients from the local herb woman or the apothecary in a nearby town.

Bess's Turkeys

Sir Walter and Johanna St. John were fond of turkey. In a series of menus dating from 15 December 1662 to 11 January 1662/23, scarcely a day passes without a turkey-based dish appearing on one of the tables at Lydiard Park. Often the household would enjoy "turkey" at one meal followed by cold turkey, turkey hash, or turkey pie at the next meal or on the next day.[14] Turkeys were introduced to Britain from the New World in the 1520s by early seafarers and explorers. By 1555 the price of the birds was already legally fixed in the London market, and by the 1570s they were being farmed. By the mid-seventeenth century, large numbers of turkeys were brought to London from the countryside for sale, and families like the St. Johns ate turkeys alongside other poultry and game birds like chickens, geese, wild ducks, pheasants, and partridges.[15] By this period turkeys were already common on Christmas tables.[16]

Given the St. Johns' partiality for turkey, it is perhaps not surprising that turkeys are mentioned in more than twenty letters within the series, and it is clear that turkeys, alongside slabs of bacon, pigs, and rabbits, were frequently sent from Lydiard to London.[17] Johanna was particularly keen to discuss the Lydiard staff's work to "prepare" turkeys for the table— that is, to fatten them up. The "turkey letters" serve as an entry into the connected areas of skill, livestock rearing, the observation of nature, and the dynamics of master-servant relationships.

Fattening turkeys was a recurrent topic for Johanna and Hardyman. The letters show that she was continually concerned with the time it took to fatten the turkeys and with the skills it required. In one letter Johanna spells out her expectation that the Lydiard staff provide the London household with at least one turkey a week.[18] This turned out to be a rather tall order. In one letter Hardyman wrote, perhaps with trepidation, that despite all their efforts the turkeys at Lydiard were still young and might not be ready in time for Christmas.[19] What quickly became apparent to Johanna, though, was that her dairymaid, Bess, lacked the skills or diligence to ensure a constant supply of turkeys. As Hannah Woolley wrote in *The Compleat Servant-Maid*, early modern dairymaids were expected not only to milk cows, churn butter, and make cheese (as their job title suggests) but also to look after poultry and other farm animals such as pigs. Woolley's guide, while offering instructions on making several kinds of cream (clotted, snow, raspberry) and cheese, merely warns dairymaids that if they "had any Fowls to fat, look to them that it may be for your credit and not your shame, when they are brought to the Table."[20] Alas, in Johanna's eyes the turkeys Bess fattened definitely fell into the latter category.

Johanna's litany of complaints began with her accusations that "Bess [was] starv[ing] them," as she had seen "many kild from the barns door better." In Johanna's mind, Bess's carelessness was forcing the family to outsource turkey procurement.[21] An exchange of letters in the winter of 1661/62 begins with provisioning but quickly transforms into a discourse about staff management. In November 1661 Johanna wrote to Hardyman requesting, as usual, more turkeys. At this point she instructs him to give Bess a crown to "incouredge her to doe it wel for we had fater when we were ther then she has sent up any sinc we came up."[22] Yet the carrot in this case seems not to have had the desired effect. Johanna's dissatisfaction with Bess's skills continued throughout the run of letters. In a later letter she wrote, "Tell bes I hope to se better Turkyes and gees of her fatting for hetherto thes we have had doth not commend her Huswifry."[23] Johanna's reference to this work as "huswifry" reminds us of the broad use of the term in this period and the multitude of skills and expertise required to be a good housewife.

As Bess's shortcomings as a dairymaid became more apparent, Johanna's instructions became more detailed and forceful. In one letter she wrote in frustration that turkeys do not take longer to grow just because she herself is not at Lydiard. She then instructs Bess to put the turkeys in the barn, keep the room clean, and "give them chang of meat."[24] How-

ever, it appears that Bess required still further instructions on her fattening technique. In a later letter Johanna admonishes Bess via Hardyman, saying she felt she had "never had so bad come to our table unles it
were when we had no corne of our own." According to Johanna, Bess was
"so afraid of fatting fowl before it is at a ful groth that she sent up 3 such
old turkyes that we could hardly tel how to eate them neither were they
fatt."[25] Furthermore, Johanna, perhaps having lost all confidence in Bess's
housewifery, is now anxious that Bess not kill her "old breeding fowl."
Johanna continued to suspect that the less-than-desirable turkeys were
a result of Bess's negligence in not providing enough food, maintaining
a clean barn, or including in the feed some "glas beaten or Brick or very
small stons to digest ther meat."[26]

Reading this series of letters, it becomes clear that while the topic is
as mundane as turkeys, there is more at stake. First, the "turkey letters"
portray a mistress with a clear idea of the minutiae of housewifery and
of rearing livestock. Johanna might live miles away from Lydiard Park,
but she was determined to control the day-to-day running of her household. She felt the only way to ensure that standards were maintained
was to issue multiple reminders and precise instructions. However, in
order to give these instructions, she herself needed to acquire in-depth
knowledge on a wide range of household duties. Elsewhere in the letters she demonstrated equally detailed knowledge on brewing beer, rearing chickens, producing medicines, and cultivating plants.[27] While much
of this knowledge could be gleaned from contemporary household and
husbandry guides, it seems likely that Johanna also had some hands-on
experience. Her use of "*my* old breeding fowl" (my emphasis) hints at a
certain attachment to a breeding scheme and to a farm animal. Johanna
might not have cleaned the turkey barn with her own hands, but it is
clear that she made routine inspections of the premises while in residence at Lydiard.

Second, this level of attention to rearing turkeys reminds us of the
multiple roles played in early modern households by livestock, game,
and other foodstuffs. Turkeys, chickens, pheasants, deer, and more were
a form of sustenance, but they were also a crucial part of the flourishing and complex gift economy. A number of studies have commented on
the importance of gifts and hospitality in forming political and social alliances, arguing that gifts were a key strategy that courtiers and gentry
used to further their own ambitions.[28] Food, in various guises, was the
most common gift.[29] As Felicity Heal has written, food often functioned

as "'little presents,' tokens of esteem, deference, or affection that are the small coin of social bonding." Their consumable nature meant that food gifts took part in a continuous dialogue of gift exchange, constructing a bond between giver and receiver. Lord Burghley, for example, advised his son to use food gifts to compliment his patrons and stay in contact with them.[30] Certain foods were offered more frequently than others, and "unusual commodities"—rare and highly seasonal—were particularly welcomed. Not surprisingly, the St. Johns fully participated in this culture of gift giving and in particular in giving venison, which—owing to the much-guarded hunting privileges held by the nobility and greater gentry—had long marked elite status. The deer park at Lydiard let the St. Johns proffer venison as a gift, showcasing their wealth, their land, and their ability to hunt.[31] As with other foods, some planning was involved in producing venison. In one letter Johanna asked Hardyman to save her the "best buck," which Sir Walter intended to present to the king. In another she requested that Hardyman give Sir Walter some warning before the venison was ready and to send up a brace of bucks, since they could "tell at any time how to bestow." Contemporaries were also not shy in requesting this prized meat for their tables; in one letter Johanna recounted how Sir Robert Pye "begged" a buck from Sir Walter and said she wished the family had a couple to give away.[32] Within the St. John household, Johanna and Sir Walter each took charge of distributing particular kinds of food gifts, and allocating venison among their friends, families, and patrons appears to have been firmly within Sir Walter's realm.[33] Johanna, on the other hand, was in charge of ensuring that smaller food gifts were sent to the right tables at the opportune times. Turkeys, it turns out, were one of her favorite gifts. As she wrote somewhat regretfully in the final "turkey letter," "I would very gladly have had som fatt Turkyes to give away."[34] Put in this context, Johanna's wrath was perhaps more about the missed opportunities for social and political networking than about dry, chewy turkey.

Rudler's Melons

Like most country estates of this period, Lydiard Park encompassed extensive gardens and grounds containing both sustenance and aesthetic sectors. These included a deer park, a rabbit warren, an area for raising poultry, pigs, and oxen, a kitchen garden, an orchard, a lake, and a series of ornamental flower gardens. Here, as with other areas of household management, Johanna kept a close eye on the goings-on in Lydiard. Seeds, detailed planting instructions, and requests for produce made frequent appearances in the letters and exchanges between Johanna and Hardyman. Their conversations on gardening echoed many of the themes introduced in the section on turkey procurement. Their discussions on cultivating melons and planting flowers, shrubs, and trees bridged issues of planning for provisions and staff management. Where Bess the dairymaid got many instructions about raising turkeys, the gardener, Rudler, was entreated to carry out Johanna's wishes in the garden. Like Bess, he received detailed and precise instructions, many demonstrating Johanna's inclination to plan ahead. She fretted about the availability of artichokes, parsnips, and "all other things that are of any long growth." Through Hardyman, she reminded Rudler time and again to ensure that these plants were ready for harvest and eating during the family's residence at Lydiard.[35]

Johanna clearly had a substantial interest in botany. In one letter she sent Rudler some seeds of the "best sort of" French melons that had arrived from Paris just the week before. She included careful directions for planting and ordering the plants in her garden. Her dedication to household matters meant she was writing these missives even when she was scarcely out of childbed.[36] On another occasion Johanna bid Rudler to try a particular experiment she had learned of from a "Dr Ingils." The experiment entailed planting a barberry or gooseberry bush the "wrong end upward," apparently to ensure that it would bear fruit "without any stones."[37] Rudler and his mistress, then, appear to have been partners in investigating nature.

In another letter she sent Rudler Portuguese melon seeds, shallot roots, cowslip roots, and the seeds of two kinds of larkspur along with careful instructions. The melons were to be planted the same way as the French melon seeds in the previous shipment, and the shallots were to be placed in the kitchen garden, with the warning that Rudler must guard them against the invading hog. Ever mindful of appearances, Johanna required that the better-looking plants be placed in the garden and the "worst" in the orchard. Additionally, she asked for sunflowers and hollyhocks "on the bank under the rails and balesters" of the garden and for stocks against the wall next to the bowling green. Finally, she implored Hardyman to ask Rudler to

git Joan and old Goodwife Woolford to git violet roots out of the woods and
Plant them in the orchard under the wal that gos from Dr Dewels garden to the

great Pond and that wall by the mil for thos violets planted in a garden wil be far blewer and better for my use.[38]

Violets were a popular ingredient in a number of household remedies, including conserve, paste, and lozenges. However, by far the most popular use was for syrup of violets—a mixture of crushed violets, water, and sugar—a common medicine to have on hand for various ailments from fevers to agues to coughs.[39] Several methods of making the syrup circulated in mid-seventeenth-century England. Some families collected multiple versions of the recipe. One of the versions in the Jacob family book, titled "To Sirup of Violets The best way coz percivalls," instructs the maker to "let [the violets] never come at any fire, nor infuse in noe water over the fire, for both those ways spoyle the colour."[40] Another contemporary recipe directs the maker to put the violet flowers and cold water into a silver mug or tankard and then place the vessel in a kettle of boiling water to heat. The maker was to take care to let the mixture stand until the "flowers are hot and the juice bee very blue" before straining the liquid.[41] Consequently it seems that Johanna was not alone in trying to ensure that her syrup of violets would be as blue as possible. Although a study of the recipes demonstrates how makers might have adjusted the methods and equipment to achieve the outcome they wanted, the St. John/Hardyman letters reveal that these efforts started even before the actual production in the stillroom, while the violets were still growing.[42]

Johanna's directives to Rudler show that she had extensive plant knowledge. In this period contemporary booksellers offered readers interested in plants a range of works from herbals to books of simples to gardening manuals.[43] While herbals and books of simples enabled early modern readers to learn about the "virtues" and uses of medicinal plants and herbs, gardening manuals like John Parkinson's *Paradisi in Sole Paradisus Terrestris* (1629), William Lawson's *A New Orchard and Garden . . . with The Country Housewifes Garden* (1618), and Samuel Gilbert's *Florist's Vade Mecum* (1683) offered information on how to cultivate all kinds of decorative flowers and shrubs, directions for shaping one's garden into ornamental knots, and month-by-month calendars to structure the planting season.[44] A number of contemporary gardening manuals were aimed at female readers, both the well-to-do and middling sorts. Women avidly read such books, using them as tools in their medical and housewifely undertakings.[45] Yet Johanna's precise instructions on how to cultivate the "best" plants also implies firsthand experience with both the ornamental areas and the kitchen gardens. Her exact instructions on where to transplant the violets demonstrate her detailed knowledge of the different growing environments in the Lydiard gardens. While it is unclear whether she attributes the enhanced blueness of the violets to soil, humidity, or sun, it is clear that such familiarity with the microenvironments in her garden could only be gained through personal observation and hands-on experience.

Many gentlewomen in this period took an interest in shaping their natural environments and, via intermediaries, in tilling the land on their estates. Margaret Boscawen, a gentlewoman living in Cornwall in the late seventeenth century, kept a small notebook in which she wrote down which roots to keep or plant in which month.[46] Perhaps because of her geographical location or her interests, Boscawen's list is largely focused on fairly common medicinal plants. Other women, though, such as Mary Somerset, shared Johanna's passion for novelties like "rare Musk Mellon seed[s]," seeking out and collecting exotic plants through her large network of correspondents. Indeed, Somerset was a key player in the trade in and exchange of exotic colonial plants and was involved in all aspects of botanizing, from collecting specimens to cultivating plants to cataloging.[47] With the managers of their large country establishments, Tregothan for Boscawen and Badminton for Somerset, it is most likely that these women also worked with a team of gardeners and helpers much as Johanna did. While their interests in horticulture join these three women, their practices and motivations probably varied. One might conjecture that whereas Somerset's interests have a strong intellectual element, Johanna's were driven by a fierce desire to maintain and improve her household's economic, social, and bodily health.[48]

Beyond Johanna's plant knowledge and spatial memory, the instructions she directed at Rudler also illustrate how gardening at Lydiard was shaped by clear objectives. It was goal oriented and collaborative across the London and Lydiard households. Each plant listed in Johanna's instructions was destined for a particular purpose, and many filled multiple roles. Some plants, like the shallots, were clearly intended for the table, in this case to make sauce. Others, such as the French and Portuguese melons, were cultivated as a novel luxury fruit that, in a gift economy where food gifts were common, would have been used both to demonstrate the family's sophisticated tastes and to build social credit. Likewise, the careful planting of flowers in the front approach of the house was designed to fashion the St. Johns' social reputation. The garden's role in creating and bolstering local "credit," as it were, can be seen in another letter where Johanna sends Rudler some seeds and some "poppy eminy" roots. Much like the French melon seeds, these were accompanied by instructions. This time, however, the seeds traveled with a word of warning. Rudler was not to "brag to much least he lose them."[49]

The letters concerning the Lydiard garden also reflect labor divisions and staff management strategies. Gardening at Lydiard was framed by collaboration across gender boundaries, and the instructions for planting violets are a case in point. Here we see Johanna telling Hardyman to instruct Rudler to ask two local women to help him transplant the flowers. This clear delineation of roles likely stemmed from a set of intertwined ideas about trust, gendered labor, expertise, and convenience. Certainly the gendering of garden work was being continually negotiated during this period, and as gardening was increasingly viewed as a skilled task requiring expertise, women's contribution began to be seen as less important.[50] We therefore might see assigning two local women to move the violets as part of this shift. However, one of the women, Goodwife Wolford, emerges elsewhere in the letter series as an expert on finding medicinal plants, and at one point she was also paid for eggs.[51] Like many such characters in the countryside, Wolford seemed to have had expertise in many areas. While it is tempting to read the separation of gardening tasks as gendered, with male servants managing and female servants doing the work, we might also consider that digging up and replanting delicate violets required a specific skill and that Goodwife Wolford and Joan were asked because of their expertise.

In addition to the minutiae of gardening, the letters add another layer to our understanding of Johanna's management style. In particular, one rather dramatic crisis point reveals that, perhaps owing to his skills and expertise, Rudler held a fairly unusual position within the St. John household. Likely to have taken place in the 1660s, the "crisis" is detailed in a run of undated letters between Johanna and Hardyman. In a move that surprised everyone, particularly his own family, Rudler ran away to London, leaving behind his wife and child and abandoning his work and personal responsibilities in the country. This act enraged the St. Johns, particularly Sir Walter, who refused to take Rudler back. The exchange of letters that charts Rudler's fate clearly shows Johanna's appreciation of his skills and her sense of obligation toward her dependents.

The exchange opened with Johanna's anxiety over Sir Walter's steadfast refusal to continue employing Rudler. Feeling otherwise, she lobbied for Rudler until Sir Walter relented so far as to allow him live "orderly and civelly with his wife and Children" at Lydiard. Additionally, if the new gardener did not "prove right," the St. Johns might renew Rudler's employment. A deep sense of obligation and responsibility for Rudler's family plainly was the driving force behind this move. Johanna implored Hardyman to "tell my neighbors I would desire they would imploy him againe for if they would have us pitty them so much as to keep him for his wife and childrens sake let them shew ther pitty in doing that for him" and also carefully instructed Hardyman not to allow anything to be "said to his disgrace becaus his credit is tender."[52]

Johanna's desire to keep her runaway gardener was due to more than just Christian duty; it also entailed her appreciation of Rudler's skills. This is evident in the rest of the letter, which was devoted to gardening matters and the division of plants in the Lydiard gardens. Rudler's dismissal triggered a negotiation and discussion over the ownership of valuable and expensive plants. Johanna was fretful because Rudler had been telling the London servants that all *her* tulips at Lydiard had died. According to the gardener, the prized flowers currently in the garden all belonged to him, and the total sum the St. Johns owed him for flowers and plants was estimated to be seven pounds. While Johanna acknowledged that Rudler had brought "many little odd things" to the garden, she argued that these were not of great value. However, she still agreed to his repaying himself by seeds and plant cuttings as long as he was monitored. Her instructions to this point were colorfully put:

> He might well repay himself by seeds and off setts out of my garden without any harm to me but I wil not have him his own carver but if it pleas god I live to com down I wil claim my own and give him what I can spar and that I think to be his but you must not let him fetch away any unles of your own knowledg he brought them out of his own garden to mine.[53]

This passage clearly shows the value early modern householders placed on both decorative and medicinal plants and reflects the lively trade in seeds and cuttings. Moreover, Rudler's skills and his contributions of seeds and plants to the garden were considered of value by his mistress. Despite his irresponsible behavior toward his family and his questionable claim to Lydiard plant materials, Johanna was still eager to employ him. One cannot but feel that poor Bess stood on quite different ground. Ultimately, it seems that everything went as Johanna had planned or intended. A few letters later, we find out that the replacement gardener was indeed a poor fit for Lydiard and its gardens and that Rudler duly regained his post. By the end of the letter series, when Johanna forwarded a set of medical instructions concerning her children's health, she readdressed the envelope this way:

> For Hardyman or if he be absent Rudler may open it and after he hath red the letter to give it to Hardyman when he coms Home[54]

Thus the once disgraced gardener so far regained his mistress's trust that he became her proxy at Lydiard. By providing luxury French melons to impress friends and patrons, intensely blue violets for making syrups and conserves, and artichokes and shallots for the table, all the while offering a splendid vista to turn the heads of visitors, the Lydiard garden was crucial to maintaining the physical, social, and economic health of the St. John household. Perhaps it is not so surprising that green-thumbed Rudler was given more leeway than other servants.

Distilling Citron Water

Health and sickness were constant themes in early modern English correspondence.[55] Many of the St. John letters began with concerned inquiries about the health of a Lydiard staff member, and others include short reports on the health of London household members and various acquaintances. A number of the letters sent to Lydiard offered medical advice and health information, both from the St. Johns themselves and from London-based physicians they knew and employed. For example, in one letter Johanna remarked that she was glad a Mrs. Weeks was recovering well but recommended, "she must often take things that purg a little to keep the matter from gathering as the Lenetive Electuary . . . the electuary to be had at any Apotheycarys."[56] When Hardyman himself was ill, Johanna wrote, "I am sory to here you have bine so ill Dr Coxe advises nothing for agues but a drink made with cardus to be drunk an Hower before and sweat and a vomit which the roots of single dafadils wil make boyld in Posset drink and taken in the morning before the fitt."[57] The few letters written by Sir Walter sometimes contain similar medical advice.[58] For example, on one occasion he recommended a local bonesetter and advised Hardyman on the signs of a well-set bone. However, in the same letter Sir Walter added, "my wife have sent him a purge," suggesting a gendered division of labor on health.[59] It seems that while both husband and wife were happy to advise on medical matters, Johanna was in charge of portioning and distributing the household's medicines. This practice is echoed elsewhere. When Sir Robert Harley (1579–1656) was ill in 1653, his son Edward wrote offering health advice gained from a Dr. Bathurst, who recommended turpentine pills. Tellingly, Edward ends his sentence by imploring Sir Robert, if it so pleases him, to commend one of his daughters to make the medicine.[60] It is unclear which one of Edward's sisters—Brilliana, Dorothy, or Margaret—received these instructions, but it is clear that in the Harleys' household, as in the St. Johns', making medicines was women's work. Thus, while men were interested in exchanging, testing, and experimenting with recipe knowledge, it seems that actually producing medicines for everyday use was assigned to teams of female workers—wives, daughters, housekeepers, and maids.

This delegation was largely because knowledge of physic, medicine production, and basic medical care were firmly seen as within the province of household management and housework more generally.[61] Women like Johanna, running large households with many dependents, tackled these tasks with the aid of a large group of household servants, male and female.[62] Johanna's and Hardyman's letters document repeated requests for distilled medicinal waters and for pounds of butter to be sent to London so Johanna could make her balsam, providing a nuanced view of how early modern gentlewomen might manage their households' health from afar. Additionally, guidance manuals for servants aspiring to work in genteel households, such as Hannah Woolley's *The Compleat Servant-Maid* (1677), often include a slew of medicinal recipes in the section designated for housekeepers, confirming that these were considered necessary knowledge for upper servants. Indeed, the image of elite women like Johanna overseeing the production of medicine and its dispensing among the household might go some way to explaining why producing medicine takes such a central place in their workdays. The case of citron water is an excellent window into these practices.

Early in the letter series, Johanna sent Hardyman a basket of citrus fruit. This note, and a recipe, accompanied the fruit:

> I have sent down 8 cittrons 12 lemmons to make som cittron water I have also sent down the receit how to stil it to Mr goram pray be you by when it is stiled I have sent down glases wherin to receive it and for wine take of that which is in

the house alredy and if loafe shuger which you have in the house wil not serve
you must send to marlburough for some white shugar candy I desire mine may
be stild in a limbeck.[63]

Johanna had several recipes in her collection for making medicines with
citrons, lemons, and oranges.[64] Citrons, a kind of large lemon, grew in
southern Europe like lemons and oranges. These citrus fruits, of course,
have many medicinal virtues, but Johanna might have been drawn to this
particular one: "The distilled water of the whole Limons, rinde and all,
drawne out by a glasse still, takes away tetters and blemishes of the skin,
and maketh the face faire and smooth."[65] The recipe for citron water in
the "Great Receit" book instructed makers to take

> Eaight of the fairest cittrons pared thin the peel of 2 orenges or 2 lemmons to
> every pel of cittron & 3 gallons of cannary steep them 10 days together close
> stoped distil it in a limbeck or comon stil which yu please.[66]

As discussed in the previous chapter, Johanna's medical recipe book
likely stayed with her in London and contained many contributions from
members of her London household and from physicians based in the city.
Thus it makes sense that she included the recipe alongside the fruit and
the instructions she sent to Lydiard. However, this recipe contained small
variations from the one inscribed in the London notebook, reflecting recipe makers' continual refinement and modification.

Much like her directives to Bess and to Rudler, Johanna also sent precise instructions for producing citron water, cautioning her staff to pare
the fruit "exceedingly thin one paring after another til you have pared it
to the pulp" and specifying that her water be distilled in a limbeck. She
also wanted to add amber, but she did not entrust this step to her servants.
Rather, since the amber was stored among her personal things, Johanna
would add this expensive ingredient herself when she was next at Lydiard. She ended her instructions by charging Hardyman to divide the used
citron peels between Mr. Goram and Elizabeth Hardyman, showing the
value placed on reuse of "leftovers."

Johanna's careful instructions remind us of the main reason householders wanted to make their own medicine: control. By making your
own, you could ensure that only the best herbs were used, leave out particular ingredients that perhaps did not agree with your constitution, and
decide exactly how sweet or alcoholic you wanted the brew. You could

control the exact method—what still to use, how finely to mash the herbs or fruit, how long to boil the mixture before distilling, and so on. You could also ensure that no one would cheat you on the expensive ingredients like amber. After all, as Johanna did, you could personally oversee that particular step. In that sense, by producing your medicines yourself or through your helpers, you could set and enforce standards. As Johanna once told Mr. Goram (through Hardyman), she "like[d] all her watters very exceeding well as posible can be."[67]

Johanna's views on homemade medicines were echoed in contemporary printed medical literature. In *The Family-Physician, and the House-Apothecary* (1676), Gideon Harvey particularly warned his readers against country apothecaries who might defraud their clients by using ingredients that were less than fresh.

In fact, Harvey's tract, according to the preface, presents recipes he selected from published sources such as *The London Dispensatory*, thereby saving his readers the hard work of reading for themselves and choosing pertinent recipes. Harvey's text is also unusual in providing the current prices for common drugs and plant materia medica to better equip consumers in their encounters with apothecaries, druggists, and herb women.[68]

In negotiations between consumers and producers of foods, as with purchasing medicines, worries about getting cheated were fairly widespread. In fact, it was not only the deceitful apothecaries Harvey described that early modern housewives needed to avoid, but also other purveyors of raw materials and goods. The short anonymous tract *The Experienced Market Man and Woman, or Profitable Instructions to All Masters and Mistresses of Families, Servants and Others* (1699) offers advice, as the title aptly phrases it, on how "To know the Goodness of all sorts of Provisions, and Prevent Being Cheated and Imposed on." Fundamental among a housewife's duties was having the savvy to not get swindled by purveyors. The skill to use smell, touch, and sight to judge the freshness and quality of various provisions was considered "creditable and commendable in Masters and Mistresses of Families, Servants and others."[69] From the richly detailed descriptions, readers learn that one can pick out young turkeys by their smooth black legs, lively eyes, and limber feet. If a turkey hen is "with Egg, she will have a fast open vent"; if not, her "vent will be close and hard."[70] Through printed manuals such as Harvey's *Family-Physician* or *The Experienced Market Man and Woman*, householders were equipped with all sorts of codified experiential knowledge to ensure smooth and economical running of their households. By combining their personal experiences and their reading, they became "smart consumers" ready to negotiate the multiple marketplaces required to best provision the household.[71]

The St. John letters also highlight the importance of season in producing medicines. The availability of many ingredients used in common recipes, such as herbs, flowers, and even butter, was limited by season. Scanning Johanna's own recipe book, we find several recipes, like "Madam Le Coques Balme or Balsame to be made in May," that had designated production times.[72] Often, instructions for making remedies using herbs and flowers included details of the best time of year for production. A good example is "Dr Boles for a Dropsey," where the maker was directed to take wormwood seed in "about August or September." Another is Dr. Allin's recipe for deafness, which required gathering a type of bluish berry in August or September.[73] Failing to plan ahead might cause extra work later on. For example, on 30 May 1661 Johanna wrote to Hardyman in a panic. She had occasion to use some old damask rose water and discovered there was none to be had in London. The apothecaries did not distill or sell the water, and the roses in the London garden were far too few and "wetted." She thus wrote to the Lydiard apothecary, Mr. Goram, who nominated "one Mrs Burg in rodburn parish at morden farm" as a possible supplier. Hardyman was to ride over there to "bespeak it all" for Johanna.[74]

This focus on seasonality extended beyond plants to animal-based ingredients. "Dr Dickinsons great Cordial Powder" required the "black tipps of crabbs claws taken in May."[75] Dr. Buttolph's remedy for a cough was also to be made in May, since this used "little white snayles that have noe shels." Presumably here the production schedule is aligned with the hatching and hibernation of snails and slugs.[76] Other compound processed ingredients were named after the months when they were ideally made. "May butter" was made in May and stored for later use.[77] Consequently, to ensure that certain favored remedies were always on hand, medicine producers spent busy springs and early summers gathering ingredients and making all sorts of balms, waters, and syrups. Aside from the availability of ingredients, many common production methods were also time consuming. Recipes for liquid medicines such as waters and syrups instructed the maker to let the herbs and spices stand or steep in the liquid carrier for hours or even days.[78] These were not medicines one could produce on demand when required. In this sense the focus on seasonality and long-term planning in the production of both foods and medicines enabled household managers to view them in parallel, with similar production cycles and workflows.

The letters make it clear that Johanna liked to be prepared for all eventualities. However, even with her careful planning the St. Johns could not always fulfill all the medical needs of their household. In one case, where Hardyman requested a remedy for "Blaks son," Johanna replied that she had "spent all that water and can make non till may." As a substitute she recommended a common cure in the period—a necklace made of peony roots that the child was to wear constantly until it fell off of its own accord.[79] Alongside wearing this necklace, every morning the patient should also take a mixture of powdered peony root and red coral washed down with a pint of black cherry water. Johanna helpfully reminded Hardyman that the peony root and red coral could be purchased from the apothecary; instructions on how to obtain the black cherry water are omitted, though, indicating that it was commonly found or already on hand.[80]

Throughout the letter series, Johanna's requests for medicines and tasks related to producing medicine were directed to various people. In most cases Johanna's instructions were brief and meant for Hardyman's wife, Elizabeth. For example, in one undated letter she wrote, "I thought your wife had stiled me some oak bud and red rose water, and would have sent it up by wolford we are very sickly both in our own house and in the towne."[81] In another letter she simply asks Elizabeth Hardyman to distill her "a qrt of mint a qrt of bawn a qrt of plantaine and 2 qrts of burag with the flowers and into that burag water put half a pnd of cinomon beatet and destil it againe and wright on it cinomon water."[82] Johanna's requests highlight the importance of domestic servants in producing food and medicine. Here Elizabeth was entrusted with compounding several distilled waters, using expensive and complex equipment, while Johanna remained in London. The absence of a written recipe for simple waters such as red rose or mint indicates that Johanna considered Elizabeth skilled in distilling and, crucially, well versed in Johanna's preferred production methods. Instructions for everyday waters like these were not included in Johanna's "great receipt" book, reminding us that the surviving recipe notebooks by no means represented all of a household's recipe knowledge. Recipes could be followed and used yet not recorded. Some might have been deemed too simple or well-known to require written codification.

In addition to distilling, Elizabeth Hardyman also took charge of preserving food. In one letter Johanna was anxious about whether the cowslips at Lydiard would "last very wel til we come down," so she wanted Elizabeth to candy them. Although Johanna did not send Elizabeth instructions for making medicinal waters (distilled or not), she did send instructions for candying, outlining the proportion of flower petals to sugar and reminding Elizabeth to "beat them very fine before any shuger is put in them."[83] While both arts were considered the responsibility of a housekeeper (Woolley also gives instructions for preserving and candying food in *The Compleat Servant-Maid*), perhaps Elizabeth was more skilled in the stillroom than in the kitchen.[84]

While Elizabeth was clearly the main producer of medicines within the St. John household, on two occasions Mr. Goram, the local apothecary, was also charged with producing medicine. In both cases documented in the letter series, it is clear that commissioning medicine outside one's household meant losing control and raised issues of trust. Johanna was keenly aware of this. In her request for citron water, she carefully instructed Hardyman, "pray be you by when it is stiled," indicating that Mr. Goram was to be trusted only under Hardyman's supervision. As detailed above, authors such as Gideon Harvey articulated common concerns about apothecaries' scruples, and there was also some shared nervousness about vendors and their wares. Yet Hardyman himself did not possess his mistress's total confidence. We might recall that Johanna was reluctant to have him rifle through her "things" to find the amber needed for this recipe. As we have seen in this chapter, the working relationship between masters and servants was complex and often tension-ridden. As we can easily imagine, household servants had ample opportunity to steal from their employers, and some certainly did so. Those landed gentry who maintained both town and country households also felt some trepidation that the staff of the vacant household would take liberties.[85] This might go some way to explain why Johanna was reluctant to give Hardyman access to her "things." In other words, we might say that Johanna entirely trusted neither Hardyman nor Goram, but directing a complex household operation from afar required certain practical decisions.

Like venison, turkeys, and plants, the medicines produced at Lydiard played a part in the complex economy of gift exchange. Conserves and distilled waters were often labor intensive to produce and required expensive and at times exotic ingredients such as fruits, sugar, and spices from far-flung places, so they had obvious value and meaning. Across early modern Europe they were exchanged and presented as gifts, as a way both to display the maker's social position and wealth and to extend friendships.[86] Like other elite women in the period, Johanna St. John also offered medical gifts to friends and family. When her nephew John Wilmot, the second Earl of Rochester, was ill, for example, she repeatedly sent both medical advice and distilled waters.[87] Thus one might see the production of homemade medicines as a means to maintain the health of the households on a number of fronts. The medicines could be administered to family and household members in times of sickness, and they could also be strategically used as gifts to garner favor, consolidate friendships, and remember oneself to one's family and kin.

Johanna also used precious medicines to build more local relationships with her staff and dependents. She clearly felt a strong responsibility for their health. We have seen how she not only inquired after their health but also dispensed her homemade medicines. Within the household, the mistress's generosity in bestowing valuable medicines on sick members must not have been lost on the servants and dependents. Johanna's dispensing medicines would have not only ensured their swift recovery but also helped her build good working and personal relationships with her staff and tenants, which were crucial to her running two households.

To manage a household remotely, Johanna relied on a large group of "upper" servants and friends based at Lydiard. Her dependence on these helpers went far beyond their producing cheese or bacon or fattening turkeys. She trusted the opinions of these men and women on a range of subjects. In one letter Hardyman and his friend were called on to ensure that Sir Walter did not fall into the snares of local naysayers.[88] The St. John children were also often sent to Lydiard to enjoy its healthful country air and the bucolic landscape. When Johanna's children fell ill, they recovered at Lydiard, often under the care of the local rector, Timothy Dewell, and his wife or the Hardymans,[89] so Johanna required trustworthy and reliable staff and friends at Lydiard to ensure their well-being. Many of the letters speak of how Johanna built up personal relationships at Lydiard. For example, in one she instructed Hardyman to pass on the two enclosed sugarloaves to the vicar's wife and to remind Sir Walter to visit Mr. Dewell while he was in residence at Lydiard, to "reward him of this pains."[90] Such reminders of her patronage and gratitude were frequently lavished, via Hardyman, on various members of the Lydiard household, local gentry, and tenants. In return Johanna expected a certain loyalty and regard.

One particular incident vividly illustrates this delicately balanced economy of regard. A furious Johanna informed Hardyman that she had

discovered her little son Oliver's wet nurse had become pregnant. Out-
raged and feeling deeply betrayed, she instructed Hardyman to "pray tel
nurs that she wil for ever forfit her credit with me if she let this child suck
after she is with child as Jug did for this [one] is weake and it would kill
it."[91] The solution to this crisis was to call in other favors from trusted
members of the Lydiard household, and the task of selecting the new wet
nurse fell to Hardyman's wife and a Mrs. Perkins. Johanna's many letters show that maintaining the "health" of a household from afar was no
easy task and required continuing cultivation and fattening, not just of the
melons and the turkeys. Staff management in an early modern household
went well beyond locating and identifying expert and reliable household
workers. It speaks of the complex relations between employer and employee, of cultures of social credit and debt, of common social codes of behavior, and of personal expectations.[92]

Conclusion

> I am so fraid of my poltry lest they should be lean.—Johanna to Hardyman,
> 14 August, 3 weeks before the visit[93]

In the early summer of 1663, rumors started among members of the St.
John household that the lord chancellor, Edward Hyde, Earl of Clarendon, would pay Lydiard a visit during the Parliament break that year. A
series of letters written in July and August reveal the preparations and
planning needed for such an event. In the first undated mention of the
visit, Johanna merely reported hearsay gained from His Lordship's servants and children. She herself placed little credence in the prospect,
since the king and queen had not yet decided on their retreat to Tun-
bridge.[94] However, despite the uncertainty, Johanna monitored His Lordship's whereabouts and his state of health over the next few weeks. A letter from 17 July stated that His Lordship was sick. Not a week later, she
informed Hardyman that she heard from Lady Clarendon that His Lordship intended to visit in August. With this news, she commanded Hardyman, "if ther be any thing to be done as to providing provision so long
before doe it."[95] Although the actual visit was not confirmed until almost
a month later, the preparations started immediately. This was probably a
good thing, since with the final confirmation came the news that His Lordship would bring with him "at least 40 in his own Train besides my Lord

Middleton who gos with him all his journey." What worried Johanna when she received this news was not the rather large group of visitors she had to house and entertain, but that she "fear[ed] becaus he is to be at so many other places in wilt[shire] that all provision wil be bespok."[96] Luckily for Johanna and Hardyman, she was a consummate organizer, and under her management the Lydiard staff worked efficiently together. Weeks before, she had already ordered that twenty to twenty-four turkeys be put aside and fattened just in case they were needed. A few days earlier, Sir Walter had already chosen the wine in London and arranged for it to be sent down. Johanna instructed Hardyman to brew the ale, hang up the green curtains, and order Smith to ensure that all the rooms in the house were scoured and dusted. A fattened hog was obtained for bacon and pork, which was apparently Hyde's favorite meat. A live ox was bought, and Hardyman was sent to a particular purveyor in "that place in Faringdon" to obtain pheasants, partridges, and quail. Additionally, he was to approach a local woman, Mrs. Church, to see if she "would give me or let me buy such as chees as that half of [the] one which she gave me last . . . for I beleve my Lord if he coms down wil like it much."[97]

Edward Hyde's 1663 visit to Lydiard highlights several important themes addressed in this chapter. First, the preparations connected with his visit, like the St. Johns' search for the king's Muscovy ducks, emphasize the link between provisioning, household management, and political and social advancement. Food production here is entangled with the fashioning of the family's social reputation and the St. Johns' efforts to maintain and consolidate their position at court and within the London political scene. The importance of food for maintaining the "health" of the family required closely monitored long-distance management of material, human, and financial resources. Within the early modern household, the notion of "health" must be taken broadly. The health of the St. John family, as demonstrated by the various interlocking vignettes in this chapter, had to be protected and maintained on a number of fronts: from keeping one's servants and workers ready to work, to ensuring that the St. John children grew up in a healthful environment, to protecting and furthering the family's interests at court by obtaining waterfowl and entertaining the lord chancellor.

Second, the detailed letters highlight the complex arrangements crucial to maintaining the "health" of the household. Owing to the seasonal availability of many ingredients, the time needed to fatten turkeys and other livestock, and the often complex and resource-heavy methods required for processing foods and medicines, planning was essential. Planning here is not only about predicting when and what one needs but also about establishing and maintaining a network of purveyors, helpers, and specialist food suppliers. Undoubtedly the diligence with which Johanna and Hardyman worked to maintain local links enabled them to call in favors for important occasions such as Hyde's visit. Preparations like fattening turkeys or noting where and from whom to obtain quail also helped Johanna and the Lydiard staff to put on a feast. The planning required to produce both medicines and foods might explain the combination of medical and culinary recipes so common in contemporary recipe collections. Undoubtedly this juxtaposition was due to the holistic view of the human body in the Galenic medical framework, but it was also a product of the multifaceted role of early modern housewives.[98]

Third, it is clear that managing large estates such as Lydiard required hands-on knowledge and lived experience in many areas. Competent and successful household managers like Johanna were comfortable in several knowledge spheres and domestic spaces, from kitchens to stillrooms to dairies to gardens. Even so, they were assisted by experts skilled in specific areas, from rearing livestock to nurturing tulip bulbs to distilling medicinal waters. The enterprise of household science required diverse sets of skills and a wide cast of characters. The St. John letters, with their at times long-winded discussions, give us a nuanced view of the "collaboration" between mistress or master and servant. As the stories about Bess and Rudler show, while defined by long-standing social and class hierarchies, these working relationships were flexible and fluid and had to be continually nurtured and negotiated. It would not be a stretch to imagine that the collaboration between men of science and their "invisible technicians" required similar care and attention.[99]

Finally, recipe knowledge—the know-how at the center of so many of these activities—played a crucial role in managing the "health" of the household. In ensuring that she had on hand the latest and "best" recipes for cake and cheese and remedies for cough, fever, ague, and infertility, Johanna was merely doing what she always did—being prepared. As housewives and household managers, women found that one key way to guarantee they were ready for any eventuality was to amass a reliable set of recipe knowledge. This meant continually seeking out, testing, and refining the best ways of doing things, from locating the best purveyor of cheese to perfecting recipes for medicines and foods.

Collecting Recipes Step-by-Step

In the spring of 1650 Edward Conway, second Viscount Conway and Killutagh (1594–1655), retired from public life. On the invitation of Algernon Percy, the tenth Earl of Northumberland, he gave up the bustle of London for the peace and quiet offered at Petworth, Northumberland's country estate in Sussex.[1] Once there, Conway kept up with the news, both political and social, through his extensive network of correspondents. Conway's nephew, Colonel Edward Harley (1689–1741), was one of the many men and women receiving Conway's letters. As he settled into life in the country, Conway encouraged Harley to begin a weekly correspondence with him. With his usual wit and humor, he wrote: "a little wind of newes would doe well to move this quiet I live in least our understandings grow mouldy."[2] For the next few years, until Conway moved abroad in 1653, uncle and nephew exchanged family gossip, philosophical and theological musings, books, bottles of cider, mead, and medicines, and recipes.[3] In fact, recipes—medicinal, culinary, and practical—were a constant feature in the letters.[4] On 29 October 1650, Conway exclaimed that on receiving Harley's letter and the "great paper," he had expected to see bundles of recipes of "Cookery or Preserving or Physick or surgery" but was disappointed to find only "Physick for the Soule."[5] A few months later, referring to their recipe exchange as "trafique in receipts," Conway chided his nephew for failing to send his share of information and the promised foods on which Harley had been practicing his skills in preserving.[6]

Aside from offering homemade foods and recipes from his own repertoire, Harley also acted as an agent for Conway, seeking out recipe knowledge that was hard to come by. In chapter 1 we saw Conway prompting Harley to use his "credit" with Lady Westmorland to obtain recipes for a face wash. Elsewhere in the letter series we find him sending Harley on missions to gather specific practical advice from local experts. For example, in a letter dated 14 October 1651, Conway directed Harley to visit a "Mr Allen the Ruler of bookes in Greene Arbor without Newgate, over against the Signe of the Parret, the second turning on the left hand," to learn how to make red ink with vermillion.[7] Conway already had an "excellent way" to make it using brazilwood, but he wanted to supplement his knowledge with an alternative method. The constant presence of recipes throughout the correspondence highlights both men's deep interest in recipe knowledge. This interest went beyond mere exchange of new information to include in-depth discussions of production methods, ingredients, and records of their personal trials of particular formulas.

In the autumn of 1651 Conway and Harley embarked on a lengthy exchange over brewing ale. This sharing of views offers a most revealing glimpse of making recipe knowledge and household science in early modern England. Despite the hundreds of collections surviving in the recipe archives, English householders were frustratingly reticent on the details of the various steps involved in producing recipe knowledge. Many collections present only recipe after recipe with little context on how they might have been used. Conway and Harley's correspondence exemplifies the multistep process through which householders assessed and wrote down recipes. Exploring this process and its epistemic consequences forms the central theme of the next two chapters, and this chapter lays out the various steps through which early modern householders put potential know-how to the test. The next chapter focuses on the central step of this knowledge codification: making and trying recipes. Taken together, the two chapters show how householders assessed recipe knowledge through a series of complex and multifaceted activities. In examining these activities, we open a window onto their trials and experiments.

I begin by introducing the central case study—a curious recipe for brewing ale that Harley sent to his uncle. I then discuss the notion of "trials on paper" and explore the varying epistemic status assigned to recipes at different stages of testing and trying. Next I turn to a detailed analysis of three recipe notebooks created by another midcentury gentleman, Sir Peter Temple. By examining Temple's use of his notebooks as paper tools to categorize recipe knowledge, I demonstrate how the multistep system of assessing recipe knowledge articulated in the Conway/Harley letters can be detected in household recipe books. The final section of this chapter opens this story to other recipe users and makers. Analyzing marginal notations, I show that Conway's, Harley's, and Temple's multistep process was commonplace in the period and the household was a site of vibrant knowledge making.

A Recipe for Brewing Ale

Conway and Harley's conversation on brewing ale began in mid-June 1651, when Conway pleaded with Harley to use his "credit" with Mary Fane, Lady Westmorland, to obtain two recipes for face wash—one prepared with ox galls and the other with myrrh. It was three months before Harley was able to answer Conway's request, and, alas, he was not the bearer of good news.[8] To Harley's request, Mary Fane replied that she had no recipes for face waters made from ox galls and myrrh other than those that Harley himself had given her. Just in case, Harley passed these two recipes along to Conway, who confirmed that these particular recipes actually originated from *his* own book and matched, word for word, instructions in his collection.[9] Perhaps anticipating his uncle's disappointment, Harley included a third recipe—one for brewing ale.

While visiting a mutual friend, Sir John Tracy (1617–1687), Harley and his host drank to Conway's health. The drink of choice was not wine but "an excellent ale." Harley so enjoyed the ale that he requested the recipe and offered it to Conway. The instructions call for the brewer to take eight bushels of malt per hogshead, boil water by itself for three hours, then pour the scalding water onto the malt and continue boiling until the water had "taken the strength" of the malt. When the mixture cooled, the maker stirred it, added yeast, and let the resulting liquid stand for a month. Although this recipe looks simple, it became the focus of a series of letters between uncle and nephew.

From the start Conway was skeptical about Tracy's ale recipe, and he pressed Harley on two points: the amount of malt used and the curious step where water was boiled by itself for three hours. To Conway this step neither improved nor worsened the "quality" of the water, and he required further persuasion before putting the recipe to the test. Despite his reservations over the ale recipe, Conway still returned the favor by sending his nephew a new and improved recipe for making mead that he was about to try. While waiting for Harley's reply to reach Sussex, Conway continued to investigate the recipe and sought out an expert—the Petworth brewer—for further advice. Agreeing with Conway on the futility of boiling water by itself, the brewer politely suggested that perhaps Harley had meant for the brewer to boil the wort, the mixture of water and malt, for three hours—fairly common in contemporary country house brewing. Conway immediately wrote again to Harley, urging his nephew to make further inquiries.[10] At this juncture the conversation turned from the practicalities of brewing to a discussion of natural philosophy. Central to the debate were contemporary theories concerning the quality and weight of water.

A letter from Conway dated 21 October 1651 implies that Harley had talked to several unnamed physicians in London, and it was there that the idea of boiling the water originated, since it was thought that this step would produce the "highest" quality water. Conway and Harley's epistolary exchange focused on what constituted the "best" water and how one might obtain it. Here uncle and nephew held differing views. Harley thought that when physicians described the lightest water as the best they were referring to the actual weight of the water. Conway, on the other hand, thought they meant the lightest water was the most easily digested. Physicians, Conway argued, "doe holde that water is not to be iudged by the balance." To further persuade his nephew, he turned to his library and offered supporting quotations from sources including Johannes Butinus's edition of *Hippocrates Aphorisms* (1625) and Louis du Gardin's *Institutiones Medicinae* (1634).

As the letters continued, notwithstanding the mounting evidence against Harley's method, Conway still retained confidence in his nephew's judgment and experience. After two months of discussion Harley finally came around to his uncle's point of view and acknowledged that "in the mutch boyling of water the thinnest and the finest parte of the water doth goe away and that you leave only the most grosse, unholsome and earthy parte." Thus he would, like his uncle, try to make ale with water that was "but little boyled."[11] But this did not signal the end of the episode for Conway, who continued to press his nephew on this matter. In the last letter about the ale recipe, Conway wrote:

> I pray let me know whether you have made good and excellent Ale by the Receipt you sent me, if you did what can you doe more, there is not anything to be beleaved in naturall things but experience, But the Brewers heare say that mutch boyling doch neither good nor hurte, but that not to boile the wort will certainly spoile the Ale and I cannot perswade any body to make it. but if you can assure me upon a triall that you have made it good I wil make triall of it.[12]

This is the last mention of the ale recipe in the Conway/Harley correspondence. Uncle and nephew continued their weekly letters for a few more years, but the conversations moved on to other matters. Conway died while traveling abroad in 1655, leaving a vast library and a trail of gossipy letters.[13] Harley lived almost fifty years more, was active in politics, and died in 1700 on his country estate of Brampton Bryan. And the ale recipe with the peculiar step of boiling the water? A survey of household recipe books suggests that it never became fully incorporated into popular domestic know-how.

Although the intricacies of this dispute lie beyond the scope of this chapter, it is significant that for Conway and Harley a discussion of making ale went beyond mere exchange of practical knowledge and was considered a topic of natural inquiry worthy of deep consideration. At stake were not just brewing a good tankard of ale but theories about the properties of water and the effects of heating it. Conway and Harley's discussions touched on a number of areas of natural inquiry.[14] Their concern over what constituted the "highest" quality water, particularly Conway's view on digestion, related to contemporary discussions by writers of diet regimens on the potential healthfulness of various kinds of potable water. Their debate could also be read alongside later discussions about boiling water and brewing beer. A few decades later two authors, the autodidact and proto-vegetarian Thomas Tryon and the alchemist William Yworth, both weighed in on the issue of boiling water for brewing.[15] Though the two were on opposite sides, the effect of boiling on water quality was a key element in their arguments. For Tryon and Yworth, boiling caused the evaporation of gases (Yworth) and spirits (Tryon), rendering the water "free and quiet ... and ... more powerful" (Yworth) or more "harsh, hard and fixed" (Tryon).[16] Harley's belief that on boiling the "thinnest and the finest parte of the water doth goe away" bears much similarity to Tryon's view.

While in-depth discussions like the Conway/Harley letters are rare survivals in the archives, it is likely that, for early modern men and women, recipes opened everyday conversations and explorations about nature, materials, and processes. As illustrated by Conway's consulting both the brewer and his own library, discussions over recipe knowledge stretched from the bookish study to the hands-on brew house, and the possible effectiveness of a particular set of know-how was considered by the household collective. Conway and Harley's lengthy discussions also demonstrate that recipes and household know-how went through a sequence of trial, assessment, and codification, and I now turn to that complex process.

Assessing Recipes on Paper and Beyond

As signaled by the continual pleas for recipes in the Conway and Harley letters, the first step in making recipe knowledge was, predictably, gathering information. This step was deeply framed by acts of sociability. From the Palmer and Bennett case studies in chapter 1 we learned that, for the enthusiastic recipe gatherer, no social occasion was unsuitable for exchanging know-how. Householders discussed health and housework over pints of ale and leisurely dinners. The transfer and exchange of recipes was shaped by the family and social networks of the knowledge gatherers. Much like the gifts of food and medicine described in earlier chapters, recipes functioned as "little presents"—small, constant reminders of affection and regard—within the complex gift economy of early modern England. In addition, as we found with the St. John household, collating know-how was part and parcel of long-term household provisioning and management. It is within this framework that Conway received Sir John Tracy's recipe for ale. The recipe is first brought to our historical actors' attention in toasting Conway's health.[17] Most certainly Tracy shared his method of brewing with Conway through Harley as a friendly gesture; perhaps he even thought of it as a "little present." It is unlikely that he expected his ale recipe to become such a topic of conversation. After all, for Tracy this was a valuable "tried and tested" method. To him, the recipe represented a code for practice—a written record of the rules that he, or his brewers, had followed to produce the particular batch of ale he shared with Harley.

For Harley and Conway, though, this recipe was assigned a slightly different epistemic status. For both, gathering the recipe was merely the first stage in the production of recipe knowledge. The next step was assessing the recipe on paper; hence the detailed discussion over its merits or faults. The letters reveal that within this "trial on paper" our historical actors relied on several sources to evaluate and verify the know-how. As the two men worked through their queries on paper, they separately consulted experts in London and at Petworth. Harley continued to speak with the unnamed physicians in the city, and Conway consulted both the learned tomes in his library and the local expert.

Conway, as far as we know, never put this recipe into action. He may have consulted his brewer about the various steps, pressed Harley for clarification, and browsed through his thousands of books to persuade Harley of his mistake, but ultimately all he promised was this: "If you can assure me upon a trial that you have made it good I wil make triall of it."[18] For Conway the ale recipe always remained on paper only. That is not to say he shied away from making up recipes. As detailed in the next section, there is evidence that he was active in trying, testing, and making medicines and drinks. Yet the ale recipe remains, at least for Conway, a mere possibility that required further consideration and "trial" on paper before any human and material resources were directed toward testing it.

If we read between the lines, despite Harley's repeated arguments for the boiling water step, the recipe's status remained somewhat ambiguous for him as well. In his letters, Conway repeatedly informed Harley that Harley's own experiences with the ale recipe were central to Conway's views. By the final letter it is clear that Harley was in the process of trying the recipe but was not yet able to endorse the recipe from his own experience and assure Conway of the validity of the production method. Perhaps Harley viewed the recipe only as an intriguing proposition, or perhaps he was still tinkering with the production method. What is certain here is that for him the recipe was not a record of practice.

For our three actors, Tracy, Harley, and Conway, the ale recipe occupied a shifting epistemic status and was at different stages of assessment, testing, and use. Within Conway's knowledge circle it was concurrently a record of tried and tested know-how and a set of instructions yet to be tried out. The epistemic state of a recipe was in flux and inherently unstable. Not only did a recipe's status shift from one actor to another as the know-how circulated, but it also was frequently redefined while in the possession of a particular actor. In October and November 1651, for Conway the ale recipe may have been just a theoretical proposition that required constant examination. However, if Harley managed to produce convincing evidence that he had "made it good," its status in Conway's eyes would shift from a mere proposal on paper to something worth trying out to, perhaps, one day being a record of how he would brew the ale himself.

The shifting epistemic state of recipes from reliable code of practice to propositions on paper, as they traveled from householder to householder, had important consequences. First, assessment, tests, and trials took on central roles in making and writing recipe knowledge. If the ale recipe is anything to go by, some householders felt the need to put many recipes through trials, whether on paper or in the stillroom. Second, within this framework the perceived value of a particular set of instructions was continually negotiated.[19] With the ale recipe, Sir John Tracy probably felt he was passing on treasured knowledge. It is doubtful that Conway felt the same way.

Making Recipe Trials

While Conway hesitated to waste resources trying the ale recipe, he certainly tested many other recipes. If we characterize his efforts to gather information and to assess recipes on paper as step two in producing household knowledge, then hands-on trial and testing typically were the third step. In fact, firsthand observation and hands-on trial and testing both occupied central places in Conway's system for determining the value and efficacy of household recipes. Recipes and practical knowledge could be hunted, gathered, and evaluated on paper, but the only way to verify that a recipe actually worked was to "make triall."

A good example of a recipe in the third stage of assessment is the mead recipe I mentioned earlier. When he sent it to Harley, Conway endorsed the set of instructions as one he believed to be good and, crucially, that he himself intended to test. Early in his stay at Petworth, Conway sent for furnaces and stills, perhaps to supplement the resources his hosts offered. Since he complained to his nephew of boredom, it seems Conway had the leisure time to test out recipes.[20] Thus, when he informed Harley that he would be testing the mead recipe, he likely meant that he, with a team of Petworth staff, intended to make the drink following the instructions provided.[21] However, at this third stage, recipes still have an ambiguous status. On one hand, they have not yet proved their worth and thus have not been assimilated into Conway's fund of practical knowledge. On the other hand, they appear to be worthy of testing by Conway's community of knowers.

Within Conway's process, this hands-on stage of trial and testing can take several guises. For medical recipes, Conway's testing and trying involved observing, directly or indirectly, the drug's effect on human subjects. For example, in one letter Conway happily informed Harley that a limewater recipe had done "great cures" at Petworth. Encouraged by these observations, Conway proposed to take the medicine himself to test its effects. He also relied on his acquaintances' experiences and observations to evaluate the efficacy of particular cures. This was most clearly shown in the case of Weston, "the man of the Melon garden," and the secret recipe for the bloody flux. In October 1651, on discovering that Harley was troubled by the ailment, a concerned Conway offered him a cure he had obtained years before from a Colonel Grantham. The cure was considered "secret" since Grantham was initially reluctant to disclose it, and it was only after much effort that Conway found out the cure was in fact powdered ginger. Now he offered it to Harley in hopes that it might alleviate his condition. In return, though, Conway asked Harley to run another errand for him—an errand that concerned the said secret powdered ginger. A while before, Conway had also shared the cure with Weston, and now he was keen to learn more about Weston's experience. That Conway was willing to go to great lengths to learn Weston's reaction to the cure shows the value he placed on the man's personal experience in assessing the recipe. Detailing his personal trial of the cure was one way Weston could repay Conway's generosity. Here the call for experiential knowledge situates Conway's practices within broader movements in early modern medicine and science, where empiricism rapidly gained in status and was increasingly used to endorse knowledge claims.[22]

Although Conway was happy to share the composition of the secret cure with his nephew, Weston was not privy to the same information. Conway ended his letter with stern instructions not to reveal the secret of the cure, since Weston knew only that he was taking a remedy for the flux and not its exact composition. Weston and Harley, though both recipients of the cure, occupied very different statuses within Conway's knowledge network. Testing blindly, Weston was never welcomed into Conway's close circle of knowers. In this instance a number of factors would elevate Harley's trustworthiness. Aside from their kinship, Conway's letters to Harley often took a pedagogical tone—an older, wiser uncle enlightening his nephew on a variety of matters. Consequently Conway might well have viewed Harley as having been trained by him. This training, paired with Harley's social standing, may have made him a different kind of witness than the melon gardener.[23] However, Harley's experiences and observations could be relied on only to a certain point. As demonstrated for brewing ale, each recipe still needed to be tested by Conway's own hands.

Conway and Harley's detailed letters highlight the instability of recipe knowledge. The epistemic status of a recipe shifted from maker to maker and from household to household. Recipe texts were records of

interrelated processes of testing, trying, codifying, and recodifying knowledge.[24] Conway and Harley's recipe-filled correspondence also sketches
out a multistep procedure for codifying knowledge that involves collecting knowledge, testing recipes, and assimilating the information into one's
personal treasury of practical know-how. If the case studies in previous
chapters emphasize the breadth and scope of early modern recipe collecting, Conway and Harley's story reminds us that recipes caught by this
wide net were put through a series of trials—on paper and in practice.

While it is unusual to have such a detailed description, it is important not to dismiss this story of Conway and Harley's brewing ale (or in
Conway's case, not brewing ale) as merely an exceptional case. The early
modern household was, more widely, a lively site of knowledge production. In the next chapter we will further examine the details of recipe trials
and how, through them, householders explored the natural and material
world around them. In the rest of this chapter I show that this multistep
system was widely adopted and can be traced in manuscript notebooks
created by other families and households.

Sir Peter Temple's Three Notebooks

In recent years note taking and notebooks in particular have come under
intense scrutiny. Scholars of learned reading and writing practices have
demonstrated the central role of "paper technologies" such as notebooks
and loose paper slips in strategies of information management and defined note taking as an epistemic activity.[25] Collating, categorizing, and
sorting reading notes required note takers to identify orders of knowledge. Parallel to the study of scholars' methods within bookish settings,
historians of science have documented and analyzed the writing of research notebooks as a means to uncover "science in the making." Studies
of the laboratory notebooks of chemists, such as F. L. Holmes's research
on Antoine Lavoisier's laboratory notebooks, enable scholars to discover
"investigative pathways."[26] By considering recipe books as "laboratory"
notebooks and "paper technologies" I take these analytical frameworks
into my examination of household recipe practices and recover the "investigative pathways" of householders. In doing so I also add the myriad
notebooks created by householders, male and female, to current conversations on information management and note taking, offering a new
space to explore the effect of paper tools in histories of knowledge. My

case study in this section centers on Sir Peter Temple (d. 1660) of Stantonbury, Buckinghamshire, and his three recipe notebooks.[27]

Temple's notebooks are all filled with medical and culinary recipes;
but, differing in form, appearance, and organization, each served a different role in his household practices.[28] The first folio-sized notebook
bears the inscription "for my dear daughter Elinor Temple" on the title
page and appears to be a gift book that Temple compiled for his daughter (MS Stowe 1077). Elinor's gift book was a carefully planned project
clearly designed from the outset. Each page was ruled and paginated, and
most of the information was written in Peter Temple's hand. The collection is well organized, with distinct sections for medical, culinary, veterinary, and household information. The medical information in particular
is arranged alphabetically by ailment and, as in many collections, blank
pages are left at the end of each letter section for inserting new information. Temple marked out the different kinds of practical household knowledge using separate tables of contents, suggesting that he saw medical
knowledge in a category distinct from general household knowledge.[29] A
list of "Cautions in the use of this Booke" follows these tables and provides background information on the workings and use of the text including his decision to encode the names of the recipe donors and key ingredients in a simple cipher.[30]

Two more volumes of recipes, one a folio-sized notebook (MS Stowe
1078) and another a small, portable duodecimo-sized notebook (MS
Stowe 1079), survive alongside Elinor's gift book in the British Library.
In contrast to Elinor's gift book, these two notebooks show little signs of
planning. Here culinary, medical, and food preservation recipes are entered alongside each other in different hands, including Peter Temple's.
The information apparently was entered chronologically. These three
manuscripts contain a number of overlaps. All contain many recipes from
the Temple, Tyrrell, and Alston families and several recipes out of Gervase Markham's *The English Hus-wife*.[31] Both MS Stowe 1078 and MS
Stowe 1079 show signs of use and bear traces of repeated reading and engagement with the practical knowledge they contain. One recipe, Dr. Winstone's remedy for a fever, is written on a loose leaf of paper tucked into
Stowe 1078 and has "entered" written on the top and bottom left corners
(fig. 3.1).[32] This recipe, with modifications, is written out in the gift book
for Elinor.

Besides Dr. Winstone's recipe, several others can be found in all three
or two of the three notebooks. Many of these carried endorsements from

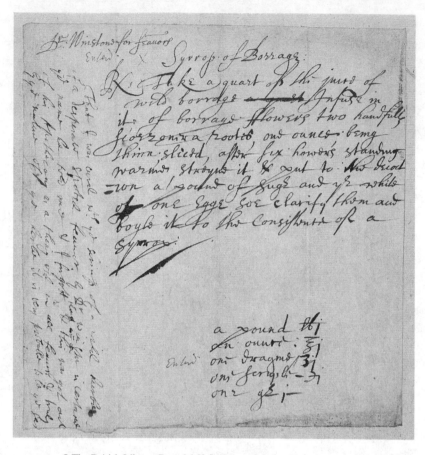

FIGURE 3.1. © The British Library Board, MS Stowe 1078, loose slip.

Temple himself and appear to be direct records of his own experiences and observations. For example, at the end of Lady Hester Alston's "fistilo water" in the gift book, Temple noted his approval: "This is a most exelant and tryed medicine, and hath prevayled when other medicins have fayled."[33]

The numerous recipes occurring across the three books, albeit in slightly different forms, indicate that Temple was copying recipes from notebook to notebook, using each to contain a specific sort of recipe. Exploring the overlapping recipes shows that the recipe knowledge the three books contain was connected in complex ways. Leaving aside Elinor's gift book, the other two notebooks, MS Stowe 1078 and Stowe MS

1079, took on distinct roles. The folio-sized Stowe MS 1078 contains culinary and medical recipes written in several hands. It probably functioned as an intermediary space, housing recipes deemed ready for testing and trial.[34] The duodecimo-sized MS Stowe 1079 is filled with recipes and likely served as Temple's *vade mecum* where recipes of unknown efficacy and status were written when and where he encountered them. Stowe MS 1078 and MS 1079 likely functioned as a set of "rough" or "waste" books where recipes collected from acquaintances or from books were written down and awaited testing.[35] If a particular item passed the test it was transferred to the gift book. The different status of these notebooks is confirmed when we see that while many recipes were transferred from the portable *vade mecum* (MS Stowe 1079) to the intermediary book (MS Stowe 1078), only a more select group were copied neatly into Elinor's gift book. Not all recipes traveled the same path through these three notebooks. Some were clearly not up to scratch. For example, next to Lady Eliza Tyrell's recipe to alleviate bloody eyes in Stowe 1078, Temple noted, "I am not satisfyd with this."[36] Not surprisingly, this recipe does not appear in the gift book. In fact, though Eliza Tyrell, probably an aunt from his wife's side of the family, donated a number of recipes to the rough notebook, only one was written out for Elinor.[37] Other recipes, like Lady Alston's fistula water, traversed the three books relatively unscathed. Still others, such as Lady Tyrell's eye water, discussed in the next chapter, accumulated additions and amendments through use and trial.

Recipe trials, whether on paper or on human bodies, are the basis of Temple's continually shifting recipes from one book to the next. His engagement with recipe knowledge reflected his constant curiosity. All three books contained little notes with questions signaled by the Latin *quaere*, a common notation in the period.[38] Examining Temple's questions suggests that, much like Conway, he used trials on paper to ascertain the potential value of a recipe before dedicating resources to making it. For example, next to a recipe said to cure convulsive fits, Temple noted the need to find out how this particular recipe differed from the version in the "dispensatory."[39] In the *vade mecum* Temple referred to the "Culpepper Dispensatory"—Nicholas Culpeper's translation of the *Pharmacopoeia Londinensis*—implying that the book was in his library.[40] He was therefore comparing his gathered recipes with an authoritative printed work. In another example, at the bottom of a recipe for internal ulcers, Temple wrote, "Quare farther into the nature of the disseases it may be proper for."[41] Given that Temple had a penchant for universal medicines, he was undoubtedly investigating whether this medicine could be used for multiple ailments.[42] Two instances see Temple returning to the recipe donor to inquire about the effects and virtues of the remedy. For example, at the end of Colonel Lovelace's version of "Sir Walter Rawleyes Pills," Temple wrote, "He sett not downe the use of these Pills but that they are Purging. Querae farther."[43] Another set of instructions for a cordial water, obtained from the nephew of the famous astrological physician Richard Napier, has inscribed "Qua[re] farther of the vertues."[44] In some other cases Temple merely wanted more information. For example, he wrote "quare further" below the main instructions in a recipe for a "Balsome to be head at Hanburg."[45] It seems likely that, like Conway, Temple was unwilling to put resources into testing these recipes while still uncertain of their value or use.

Temple also took a large number of recipes to the second stage—making and trying them. Throughout Elinor's gift book, Temple constantly reminded future readers that the cures it contained had been proved by experience and trial. Recommendations such as "this I have often proved with good success," "this is it I used," or "experientia docet" are attached to a select few of the recipes.[46] Temple's privileging of firsthand experience in assessing cures was a widespread practice in this period. For example, Alisha Rankin, in her study of sixteenth-century German courtly recipe collections, has shown that German princesses and courtiers also vouched for many cures by citing their own direct involvement either by witnessing the cure or by testing it on their own bodies. This emphasis on experiential knowledge, she argues, was a "key source of medical authority for a large swath of the population," including learned physicians, artisans, and medical practitioners of all stripes.[47]

To highlight the tried and tested recipes in the gift book, Temple compiled an additional table of contents titled "A Particular Table of proved and Experimented Receits."[48] The table listed alphabetically thirty-four recipes for medical remedies that had been tested. Throughout the text, Temple signaled these recipes by writing P.E. in the margin. Here P.E. likely stands for *probatum est*, "it is proved," a phrase denoting efficacy commonly used in medieval and early modern recipe collections.[49] The number of recipes annotated with P.E. is higher than the number listed in the additional table of contents, indicating that even a meticulous compiler like Temple did not always keep totally up to date.

In compiling the list and marking particular recipes P.E., Temple relied on both his own experiences with the cure and the experience of the recipe donors. One instance is Dr. Winston's cure for a "spotted or other feaver," recorded with a backstory recounting Temple's own illnesses. During one bout of suffering from a spotted fever, Temple was cured by a "syrup of a wild hearb (which grew common in every pasture)" prescribed by Dr. Winston. Though Dr. Winston helpfully disclosed the name of the herb, Temple forgot it and had to ask for help from Winston's apothecary, who aided him by disclosing all the common herbs Winston prescribed for such illnesses. After making a trial, Temple was almost certain the apothecary had named the "right" herb. Temple particularly appreciated the low cost of the drug, as he wrote in Elinor's gift book: "What was most remarkable was my physicke came not but to 24d all the time."[50] In particular instances, Temple would endorse a medicine as efficacious but state that he preferred another drug. For example, under a recipe for "lozenges for a cold," Temple wrote P.E. and "this have I often seen prevayle yet give me the old medicines."[51]

Many of these personally endorsed recipes in Elinor's gift book also appear to be direct evidence of Temple's own recipe trials. These include a recipe for a weak back, labeled "my own," to which Temple appended, "but this alone (to my knowledge) hath (in divers) [ways] wrought rare effects. And I feele the powerfull effects and benefitt of it above 7 year after."[52] Another explicit statement of Temple's own experience comes with a recipe that was filed under K for kidneys but was also labeled "An universall medicine for divers Greefs and maladys called Solus omnium ungentorum."[53] Like many others, this recipe was recorded in all three notebooks and was transferred with only a few additions. One of the additions, written only in the version in the gift book, advised makers to shape the resulting mixture into rolls and to be sure to grease their hands to keep it from sticking. Another addition, an endorsement of the recipe (in two slightly different formulations), can be found in both the *vade mecum* and the gift book. Although the additions in the *vade mecum* were written in a darker ink, they are copied in the same ink as the main text in the gift book. The endorsement reads:

> To my knowledge, Sir Edw Tyrill used this as an universall medicine for the disseases above say'd (*and was his Faithfull frend on all occasions for the gout, which never came more where it was once applyd*) and I have often had experience in my owne family (both in man and beast) on occasions, till my balsome put it out of request with mee. *Yet hath it not lost its virtue*, nor is to be rejected, P: T.[54]

This lengthy endorsement confirms Temple's use of the gift book to collate tried, tested, and proven recipes and illustrates the importance he put on personal experience and observation of the cure. In this particular case, while the endorsement reveals that Temple had dispensed the drug and witnessed its effects, the recommendation that makers grease their hands before handling the mixture hints at his direct experience with making it.

Aside from basing his recommendations on personal experiences and hands-on trials, like Conway, Temple also relied on the endorsements of others. In some recipes, such as Sir Kenelm Digby's recipe to hasten labor, which Temple was obviously not able to test on himself, he provided an anecdote to support its efficacy. He noted at the end of the recipe, "if I mistake not the Lady Digby sent my wife some of this in some extreamity and she had present speede."[55] Another example of a recipe donor's recalling the success the remedy had on another person is Mrs. Hinton's recommendation for the "hot goute," which "she hath often proved on her Husband who was much troubled but let him be never so ill over night he could walke in the morning and is not now troubled soe much with it as formerly."[56] Other recipes came to Temple with endorsements backed by the donors' personal experience. For example, a recipe for an excellent pomatum for the face, attributed to Lady Forster, came with the claim "This was tryed by a friend of mine who sayes tis the only one that ever she saw that did not yellow the skin and that tis as good as costly."[57] Another recipe for "consumption for the lungs decay," given by Mr. Clearke, is supported by "[this] prevayles where all other remeadys have fayled as sayth the Author he hath experiment[ed] upon his owne body."[58]

Like Conway and Harley's letters, Peter Temple's three recipe notebooks demonstrate the multistep system that many householders used to create their treasuries of health. Whereas Conway described his system of recipe assessment in his letters, Temple offers us a hint of his practices in his multiple notebooks. He first cast a wide net for information that he wrote, depending on his location, one presumes, either into his *vade mecum* or directly into the intermediary book. There the know-how is still under review, in the same way that Conway questioned the ale recipe Harley offered. At this stage Temple repeatedly verified his collected recipe knowledge by consulting printed books and recipe donors. After this, much as Conway tested the mead and limewater recipes, in the intermediary stage Temple then produced and tried the recipes himself. After multiple tries, when he was satisfied that it was worth keeping, he accepted the recipe into his canon by writing it in the book he was creating for his daughter.

Studying Temple's three books together reveals a certain flexibility and fluidity in making recipe knowledge. Many of the recipes transferred from book to book show changes as Temple tried, tested, and modified them to suit his needs. The crossovers between the books are also not entirely consistent. While many recipes existed in all three books, some, such as Lady Forster's pomatum, are recorded only in the *vade mecum* and the gift book, and others, like many of the eye remedies taken from Lady Eliza Tyrell, were written only in the intermediary book and the gift book.[59] In certain cases recipes were transferred into the gift book with queries unanswered. An example is Eliza Tyrell's recipe to address the "falling of the palatt of the mouth." The instructions, first appearing in the *vade mecum*, do not specify the quantity of herbs and other ingredients required. Temple flagged this as an issue in the intermediary book, but the same recipe, with no additional information, is written into the gift book.[60] Codifying recipe knowledge was a messy business.

While the exact relation between the three notebooks, in both temporal and epistemic terms, remains murky to anyone other than Temple himself, it is clear that he used his notebooks to differentiate recipes in different stages of assessment. As he transferred recipes through the steps of knowledge production, each notebook played a distinct role within the family's strategy for collecting recipes. Viewed together, the three notebooks emphasize that although compilers might gather information from a variety of recipe donors, they chose only a select few for use. Temple's use of a *vade mecum*, an intermediary book, and a neat copy (the gift book) highlights the strict selection recipes underwent before they were accepted into the family's daily practice. Within the scheme of knowledge production, the physical notebooks served multiple purposes. The gift or neat book delineated a set of treasured medical knowledge, experienced and tested by Temple himself. The *vade mecum* and the intermediary book functioned as a mode of storage, an in-between or liminal stage where information awaited testing and evaluating before graduating, as it were, to the family treasury.

Temple's use of paper notebooks to hold recipes of different epistemic status and value speaks to the continuities between learned and vernacular practices of note taking, information management, and knowledge production. Yet the location of his undertakings within the household pushes us to consider his efforts alongside other kinds of domestic and personal writings. Recent studies have argued that early modern life writings, whether in account books or in the margins and blank leaves of

almanacs or commonplace books, also were often reworked and rewritten.[61] Temple's three notebooks, and others like them, suggest that other
householders treated their recipe writings in much the same way, reminding us of the close connection between the various textual genres and
knowledge practices in early modern homes. I will return to these themes
of life writing, family history, and recipes in my final chapter.

Virtual Rough and Neat Books

Although the case of the three Temple notebooks, with the survival of the
waste and neat books, is unusual, it was not uncommon for householders
to own multiple recipe notebooks and use them concurrently. Some, like
Johanna St. John and Frances Catchmay, used different books to separate
culinary recipes and medical knowledge.[62] Archdale Palmer also worked
with multiple books. Information on cleaning horses' harnesses and other
equipment comes with the annotation "see in another receit book p.
147."[63] In these cases householders used notebooks to categorize medical,
culinary, and veterinary knowledge. Across the recipe archive there are
also instances of householders' using multiple notebooks in their trying
and testing. One such book is the late seventeenth-century recipe notebook associated with Elizabeth Godfrey.[64]

Now in the Wellcome Library, the vellum-bound quarto-sized notebook with gilt decorations is well worn. Godfrey and subsequent readers
and users left many marginal marks recording their engagement with its
information. A series of annotations shows that the users of the Godfrey
volume actively transferred tried and tested recipes to another depository—a "green book." Two recipes for bread pudding (one baked and another boiled) were considered worthy of transfer into the green book and
marked with "this I have write in my green book"[65] (fig. 3.2). Like Temple,
the users of the Godfrey notebook did not find all recipes worth keeping.
Recipes for a variety of baked goods (biscuits, Mrs. Bradford's seed cakes,
and cheese cakes) were rejected with "not to be write."[66]

The transfer of approved recipes into the "green book" was only one
method the readers of the Godfrey notebook used to document their
household practices. They adopted a range of other techniques to select, accept, and reject recorded practical knowledge. Useful recipes were
marked with a heart-shaped symbol or a check mark, and rejected recipes
were crossed out.[67] Recipes annotated with the heart include a recipe for

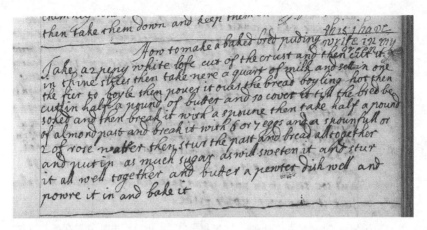

FIGURE 3.2. Wellcome Library, London, MS Western 2535, fol. 45.

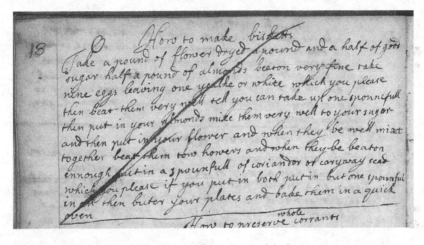

FIGURE 3.3. Wellcome Library, London, MS Western 2535, fol. 18.

the "Lady Wootton's cake," one for "biskett" bread, and one for fritters.[68] Rejected recipes include one for candying angelica, another for liver pudding "Mrs Walkers way," and a third for rice pudding given to the compiler by "my cousin cook."[69] One recipe for biscuits was both marked by the heart and crossed out, suggesting that the readers selected the recipe, tested it, and found it wanting[70] (fig. 3.3). The readers of the Godfrey manuscript, then, not only carried out testing as Temple and Conway did but also used their multistep collection procedure within their households.

The Temple and Godfrey notebooks provide rare but definite evidence that neat and rough books could be used to organize, evaluate, and sort new information. In other cases the transfer of selected information from one knowledge repository to another was either not carried out physically or not overtly expressed. However, that is not to say contemporaries did not employ similar strategies. Many of the surviving recipe notebooks have textual traces that suggest this culture of trial and testing. Like the readers of the Godfrey notebook, these other compilers left their traces through common reader's marks.[71] Examples of these practices abound in surviving manuscripts. The book owned by Elizabeth Okeover Adderley (1644–1721) offers a particularly rich example.

The book in question, now housed in the Wellcome Library, is a quarto-sized notebook bound in limp gilt-stamped vellum covers.[72] Inside, recipes have been entered in a number of hands, and one writer identifies herself as "Eliz. Okeover now Adderley."[73] This notebook is one of two associated with the extended Okeover family from the area around the Staffordshire and Derbyshire border.[74] The two Okeover notebooks are linked by a number of overlapping recipes that an anonymous household member at Okeover Hall copied verbatim from the earlier Okeover manuscript to the book now associated with Elizabeth Okeover Adderley. Both Okeover notebooks were used by members of the households and provide glimpses of the medical practices of the two Okeover families.[75] The readers/users of the Okeover/Adderley family book were avid annotators and commentators and, like Elizabeth Godfrey, used various strategies to record their practices of testing recipes, selecting information, and making medicines.

Like many other early modern families, the Okeover/Adderleys accumulated a large corpus of medicinal and culinary recipes. Their collecting pattern largely follows the patterns of recipe collecting described in chapter 1. Like many similar books in the period, the Okeover/Adderley recipe book was written in multiple hands, with recipes entered chronologically. It contained a section of recipes copied from an existing family book, perhaps a "starter" collection, and readers and users are invited to find particular information using the detailed alphabetical index at the end of the volume. Once the information was entered, like the Temple and Godfrey families, the Okeover/Adderleys began selecting and testing recipes. The initial selections made by the family are noted with a plus sign in the margin. These marks can be seen throughout the volume, and their large number suggests they record the family's interest in particular recipes rather than their actual experience with them—that is, these are probably selection marks. The family tended to be vocal about tried and tested recipes, which were actively marked throughout the book and in the index. For example, next to a title "ffor a Blast a most exelent oyntment" in the index, a user wrote "tried & is good." Someone also wrote "good" next to two entries "for the Ague."[76] Perhaps tired of writing out the entire word, another reader marked a selection of recipes in the text with the letter *g*. Several of these entries, such as the recipe "For a Pearle in the eye," carry both the initial plus sign and the letter *g*, hinting that these recipes were first selected, then tested, and proved to be worthy.[77] Marks like these locate the Okeover/Adderleys' engagement with recipe knowledge alongside the multistep schemes adopted by Conway, Temple, and others featured in this chapter.

To further confirm the place of testing in their schemes of recipe knowledge making, the Okeover/Adderleys also used more precise language—"proved" and "cured"—to describe their experiences with particular remedies. The recipe for "A caudle for a woman with a fflux in childbed Dr Ch:" is marked out with a plus and a *g*. In the margin is written, "very good for a lask Proved in childe Bed or any other time."[78] The term proved is also written as "I: pro" in the margin next to the recipe for "An Eye water against a hot cause E:G." and another "For to prevent a miscarriage."[79] The Okeover/Adderleys also recorded certain recipes as having cured specific bouts of sickness, signaling particular family members' successful experiences. For example, a recipe for the rickets is endorsed with "this cured my Brother."[80] In the other instance, next to a pair of recipes given by Dr. Dakins, a bolus and a purge addressing "gripes in the guts and staies loosnes," is this note: "it cured my mother many voyolent fits takeing a dose of the bolus the same night & 2 nights aftter a small quantie of it as occation requiers."[81]

As the family gained experience with particular recipes and knowhow, some became their go-to remedies for particular ailments. These are marked with annotations such as "this I make Eliz: Okeover now Adderley,"[82] or "that my mother always maketh to use,"[83] or "this is the salve I allwais make." The final statement was written next to a recipe for "fflos Unguenttorum: a most excelent salve." Additional marginal annotations connected with this recipe demonstrate how the Okeovers, like many other early modern families, modified the recipe to suit their own needs. As shown in figure 3.4, the main ingredient of the salve is rosin. In the original recipe, only pure rosin is required. Here rosin is likely to refer

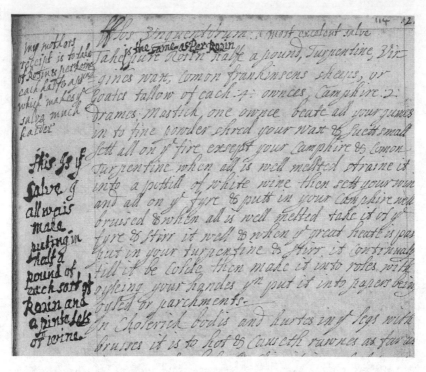

FIGURE 3.4. Wellcome Library, London, MS Western 3712, fol. 114v.

to a solid resin produced as a by-product in the distillation of oil of turpentine.[84] Two marginal notes, though, indicate that the later users of the book tended to follow a modified version of this recipe. The first note, likely written by Elizabeth Okeover Adderley, reads: "my mothers receipt is to take of Rosin & per Rosin each halfe a pound which make the salve much harder."[85] The second note states: "this is that salve I allwais make putting in half a pound of each sort of Rosin and a pinte less of wine."[86] Okeover Adderley further records her own experiences with the cure in the index. Next to the recipe's entry there she instructed future readers that this was also called "that soft yellow salve" and signaled her own use of it by adding her initials in the margin: "E. O."[87]

Elizabeth also signs her initials next to several other recipes in the index, signaling to future readers that she was also fond of two recipes for a black salve.[88] Elizabeth Okeover Adderley's use of "I allwais make" in connection with the yellow salve shows that these particular drugs were commonly used and continually produced in their household. There were,

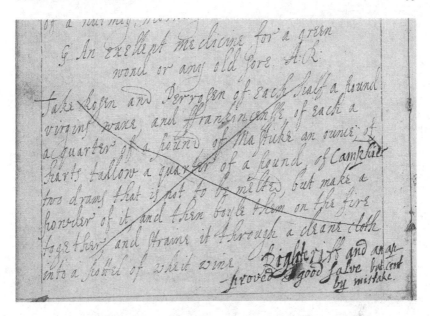

FIGURE 3.5. Wellcome Library, London, MS Western 3712, fol. 6v.

of course, other medicines in the book that were made up only infrequently. For example, a user, likely Elizabeth Okeover Adderley, wrote, "it I used once or twice" next to an index entry for a remedy "for a weakness in the Back so that one cannot sit up."[89]

But what of recipes that went through the testing and did not make the grade? It seems likely that, like the readers of Elizabeth Godfrey's book, the Okeover/Adderley family just crossed them out. Crossed-out recipes are dotted throughout the notebook. With the recipe for "An exellent medicine for a green wound or any old sore A:R:," the users left clear indications of why recipes were crossed out. This one is another version of the fflos Unguenttorum/yellow salve recipe discussed above. In this iteration someone has scrawled two large crosses over the recipe—a common sign of rejection. However, the same person amends (in a different ink): "Right ritt and an approved good salve but crost by mistake" (fig. 3.5). To confirm that the recipe was actually approved, a g was also written next to the title.[90] Accompanying this recipe is a loose slip of paper scribbled with what appears to be a price list of the ingredients used in the salve, with a note at the bottom: "Some Reseits In this Book by mistakes was writ twice over; so one of each is crost But this harder yellow salve should not have bine crost."[91] The long-forgotten notes on these scraps of paper thus provide a clear explanation for why early modern householders might have crossed out recipes in their books: because the recipe was transcribed incorrectly, there were duplicates in the book, or it did not prove efficacious.

While on first glance the Okeover/Adderley family book may seem a mere mishmash of recipes recorded in no particular order, analysis reveals that the family neatly grouped their tidbits of household knowledge in categories, each representing a stage within an established evaluation process. Recipes were separated into those written down for interest, those awaiting testing, those that were approved, those that had demonstrated their worth by particular cures, those accepted into the family's regular medical repertoire, and those rejected outright. Like the other compilers discussed in this chapter, the family created their recipe book in three stages. However, unlike Temple and Godfrey, the Okeover/Adderley family did not use waste and neat books and thus did not transfer selected knowledge from one physical depository to another. Rather, they used annotations—the plus sign, the letter *g*, and other symbols or phrases—to create virtual waste and neat books. For the various users of the Okeover/Adderley family book, the status assigned to particular recipes is immediately clear from the signs, marginal statements, and cross-outs. It is as if the knowledge it contains is separated into several levels of reliability, with each level clearly delineated.

The Okeover/Adderley family's use of annotations and marginal statements to differentiate categories of recipes was common among compilers. More often than not, early modern recipe books are marked up, and compilers used their own customized sets of signs and phrases to classify recipes and their stock of household knowledge. Many of the family books I feature here are heavily marked up. For example, the books of the Fairfax, Johnson, and Fanshawe families, discussed in other chapters, were all marked up with marginal signs and cross-outs. In particular, just like the Okeover/Adderleys, the Johnsons devised clear categories and used a check mark and "good" to signify the selection and approval stages. Archdale Palmer and Henry Fairfax used traditional marks like the manicule (a pointing hand), and Johanna St. John used *X*s to cross out recipes and letters of the alphabet to categorize the information.[92] Finally, Palmer and the users of both the Johnson and Fanshawe family books crossed out unwanted recipes.[93]

Some compilers, such as Mary Grosvenor, daughter of Sir Richard Grosvenor of Eaton Hall, Cheshire, placed such importance on these se-

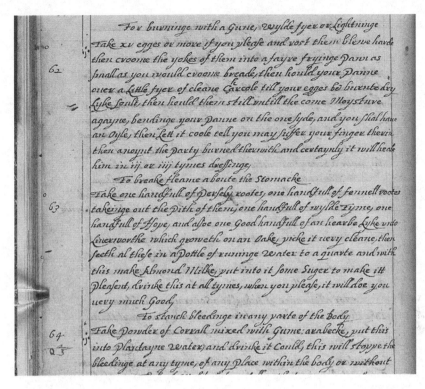

FIGURE 3.6. © The British Library Board, MS Sloane 3235, fol. 15v.

lection marks that they created a designated space for them.[94] The pages of Grosvenor's recipe book were all ruled with a narrow column on the left side. In this area Grosvenor or other users added a range of marks to select and to record uses and trials. The marks are first encountered in the lengthy table of contents, where just over half of the 264 recipes listed were annotated with a trefoil symbol.[95] Here it is likely that the trefoil was used to denote recipes of particular interest. Within the body of the book, additional annotations were placed next to a number of recipes. For example, recipes 62, 63, and 64 on folio 15v are all marked with a range of selection marks (fig. 3.6). Recipe 62, "For burninge with a gune, wyldefyer or Lightninge," is marked with a trefoil, a small circle, and two diagonal lines. Recipe 64, "To staunch bleeding in any parte of the Body," is marked with a dot, a trefoil, and a manicule. These symbols were used consistently throughout the volume and were likely part of a system for recording instances of use or testing. Alas, Grosvenor did not leave instructions on how to decipher these signs. One possibility is that each of the signs was attached to particular family or household members and used to record their experiences with individual remedies. Another possibility is that each sign marked a particular stage of testing. In both scenarios, this system of signs both documented use and trial and informed future users about the household's experiences with the remedy.

Grosvenor was not alone in devising a complex system of annotations. Another example is the work of the users reading and annotating MS 84 in the Exeter College Library, Oxford. Here the anonymous compilers marked out recipes by placing a plus sign in the margin next to the title and highlighted the recipe's success by writing "this is good" in the margin.[96] In addition, they created two external indexes on separate pieces of paper.[97] These indexes are heavily marked up, with some recipe titles crossed out entirely and # placed next to other titles. As with Grosvenor's book, it is now difficult to discern the exact meaning of these cross-outs and symbols. In this case the marks in the index do not match the annotations in the main text. One possible explanation is that the annotations in the two indexes and in the main text represent readings of the collection on separate occasions by three individuals. The markings then represent the readers' own engagement with the text and their ideas of the recipes' efficacy and usefulness.

Conclusion

From Edward Conway and Edward Harley's debate about an ale recipe to Peter Temple's three notebooks to Elizabeth Godfrey's and Elizabeth Adderley's scrawled-over recipe books, the case studies presented in this chapter demonstrate that early modern English recipe collections followed a fairly uniform multistep process of codifying knowledge framed by acts of sociability. This process began with the exchange of know-how, often during social occasions, followed by verifying the newly acquired knowledge. At this stage recipe collectors might turn to local experts, such as the brewer, or to their own libraries to consult texts by both contemporary and ancient authorities. Also at this stage they might return to the recipe donors with queries, creating further opportunities for social interaction. Because of the expense in time and money, only after success seemed possible did householders channel their energy and resources to making and testing the new recipe, and only after extensive testing and often modifying did that recipe enter the household's daily practice. This procedure challenges conceptions that householders or recipe compilers were just passive recipients of stagnant knowledge. Rather, they were continually engaged in critiquing and rewriting recipes.

Within this continual assessment and reassessment, traceability of knowledge was crucial. In many of the cases I discuss, spelling out the provenance for particular recipes enabled compilers to return to the source with queries. Moreover, carefully noting the original and intermediary recipe donors let them assign responsibility, accountability, and ownership to specific sets of recipe knowledge. Just as Conway returned to Harley time and time again to query the finer points of the ale recipe and Temple frequently wrote "quare" in the margins of his notebook, it is likely that other householders followed recipe donations or gifts in a similar fashion. Additionally, as recipes underwent multistep assessment and were tested in various ways, it was not only the actual cure that was on trial but also the reliability and trustworthiness of the source. Thus the success of the cure would be attributed to everyone named. If the cure proved to be a failure, however, the reputation of all named would be affected.

Consequently, recipes in manuscript were flexible texts where the boundaries between producer and user and author and reader were fluid and constantly negotiated. After being collated, tested, and modified, the recipe adopted into a particular collection and again released into the recipe exchange circuit was in effect a new recipe. It might resemble the original, but it likely used slightly different ingredients or maybe had an extra step or two in the production method, or it might have a slight change in title and a different author assigned. In this way seemingly thousands of fairly similar yet different recipes circulated in early modern England. The recipe, as text and as knowledge, is particularly adaptable and open-ended.

Finally, the model of making recipe knowledge I put forward in this chapter places reading and writing practices firmly alongside hands-on trials as pathways for gathering, assessing, and recording knowledge. Libraries and experts worked together to help Conway verify the ale recipe. In calling on both books and experts, he joined communities of knowledge seekers in many settings, from learned societies and academies to artisanal workshops to the streets of London.[98] The common notation practices and paper tools usually associated with learned reading and knowledge making also were key in making recipe knowledge. Rough

and neat books, practices of writing and rewriting, and marginal annotations were all employed by householders to categorize and sort recipe
knowledge. They used these tools to separate various kinds of household
knowledge—medical, culinary, and practical—and also to signal recipes'
epistemic status. The site for making recipe knowledge might be new, but
the methods and tools are familiar to historians of science and knowledge.

This focus on reading and writing practices emphasizes the record-
keeping function of household recipe books, encouraging us to view statements of personal experience or observation of cures within early modern
cultures not only of knowing but also of writing. Recent studies have outlined how, from the sixteenth century onward, experiential knowledge
was seen as increasingly important within medical and scientific discourse
across settings.[99] Through our study of the St. Johns' adventures in rearing
poultry and in gardening, chapter 2 illustrated how observing the natural
world was a normal part of a household's daily life. As I discuss further in
chapter 4, householders used their recipe trials to investigate a range of
processes, from assessing production techniques and equipment to understanding material changes to evaluating the effects of particular cures on
the human body. In stressing firsthand observations, personal experiences,
and hands-on trials in assessing recipe knowledge, English householders
certainly participated in larger conversations about experiential knowledge and cultures of "empiricism" framing the evaluation and making of
all kinds of natural knowledge. Concurrently, however, in writing down
their own experiences with recipes, they were also participating in other
practices of record keeping such as accounting or life writing.[100] As we will
explore in chapter 5, for early modern householders writing recipes was
deeply tied to writing family histories and recording events in their own
lives.

Recipe Trials in the Early Modern Household

In May 1658 Sir Edward Dering (1625–1684) made an entry in his notebook headed "For the cramp & paine in the joints."[1] The entry begins by offering a recipe for making an herb- and butter-based ointment, then details how Dering tried this medicine on a certain "Boughton of Pluckley" who was so afflicted by pain in all his limbs that he was confined to his bed. According to Dering, after only two applications of this ointment, Boughton recovered and was "well."[2] Not one to rest on his laurels, Dering concluded his trial with this thought: "But this I suppose to[o] noble and to[o] soudain an effect for those simples to produce: probably his disease was then declining."[3] This was not the only trial Dering conducted; his commonplace book is filled with recipes and with his observations of various tests and trials.

When his daughter Betty burned her left eyebrow and part of her forehead while trying to snatch a candle from her brothers, for example, Dering wrote:

> we laid to it Unguentum Album which put her to a great deale of paine: after that diaculum: and then snow water: all of which we found apparently did her a great deale of hurt: the Diaculum drawing and the unguentum and snow water being extreme cold and therefore repelling, had made run a great deale further and inflammed much more than it was: after that we laid to it only sallet ole beaten in water which fetched out the fire and gently supplying of it asswages the paine.[4]

This detailed account offers a compassionate portrayal of Betty's suffering and Dering's concern for his child's well-being. In fact, the health of the Dering family and dependents was a common topic in the notebook. Directly above the entry on Betty's burns, for example, is an entry on how her brother Heneage, previously suffering from a cough and head ailment, had come down with the measles.[5] And elsewhere in the same notebook Dering records his trying a medicine for gout on his bailiff, James Honeywood, and also gives a day-by-day description of Honeywood's final illness and death.[6] These entries show how the Derings reacted to accidents and sicknesses in their household and illustrate their testing, trying, and assessing of recipe knowledge.

The description of Betty's burns, in particular, shows the thought processes underlying household medical efforts. As concerned parents, the Derings tried medicaments until they found one that assuaged Betty's pain. This was not straightforward use of a single medicine to alleviate a specific ailment but a continuous reading, assessment, and reassessment of the effects various treatments had on Betty's body. Other entries in the notebook indicate that such speculation was not unusual for the Derings. Earlier in the same year, for example, driven by his fear of kidney stones and toothaches, Edward designed and made *aqua nephritica* and a distilled water to clean his teeth. Here he carefully noted that he "avoided all those Diareticks which being of hot and sharpe parts provoke urine rather then breake the stone."[7] These trials were not only a search for medicines that worked; they investigated how materials interacted with the human body, in an effort to further understand the natural world.

In chapter 3 we saw how early modern men and women put potential know-how through a multistep evaluation before assimilating new recipes into their treasuries of household knowledge. Here I take an in-depth look at one particular step within this process of codifying knowledge: recipe trials. Curiosity and the necessity for trustworthy know-how drove men and women to continually put their recipes to the test. Just as the drug testing by learned physicians across Europe focused on a wide range of issues, including drug composition, production methods, and the effects of materia medica on the human body, householders geared their testing toward an equally broad end. In valuing recipe trials, they participated in the pan-European attempts to try out both drugs and practical know-how.[8] Building on the arguments presented in previous chapters, I contend that the domestic setting brought with it a particular set of considerations in determining expertise, authority, and value. Investigating recipe trials in the household thus offers insight into methods of assessment and testing in a new spatial and social context and gives us a rare glimpse of procedures outside academic institutions. Within the household, recipe trials were crucially shaped by notions of sociability. The earlier case studies on the Brockman, Palmer, and Bennett families illustrated the close ties between collating recipe knowledge and social practices. In their eagerness to gather new recipes and forge new alliances, many compilers took an open approach. This openness meant that the information they collected had to be tested, experienced, and customized to suit the needs of individual households.

To explore these themes further, this chapter shifts the focus from the body of the texts to notes in the margins and scribbled between the lines. The central focus of this chapter is household recipe trials. I begin by examining the language and terms householders used to describe their testing of recipe knowledge, then I turn to the myriad ways householders tested a variety of issues as they followed recipes and made up medicines. After that I discuss "recipe salvage"—how and why compilers might attempt to save treasured recipes. Finally, I analyze the effects of rewriting, modifying, and customizing recipe knowledge.

The Language of Testing

Early modern householders recorded and discussed their testing and trying of recipe knowledge in a number of ways. In the previous chapter we learned how they used signs and marks such as X or g or a check mark in the margins to note tried and tested recipes or to record their experience with particular know-how. They also wrote short statements variously describing their endeavors as trying, experiencing, proving, and experimenting. Common phrases used include "this I have often proved with good success," "this is it I used," *experientia docet*, "experimented," to make "tryall," or the well-known efficacy phrase *probatum est*.[9] This language permeated both printed and manuscript recipe books and was used to signal experiential knowledge, hands-on trials, and personal endorsement and approval. Focusing on one writer—Johanna St. John—this section explores the various terms used to describe engagements with recipe knowledge.

The verb "to prove," appearing as "approved" or "proved" or as the Latin phrase *probatum est*, was perhaps the most common term used in both print and manuscript.[10] Many collections were touted as containing "approved" remedies. In print we might think of Alexander Read's *Most*

Excellent and Approved Medicines and Remedies for Most Diseases and Maladies Incident to Man's Body (1652). In manuscript, we find collections with titles like "A Book of such Medicines as have been approved by the speciall practize of Mrs Corlyon" or "A collection of sundrie approved receipts in phisike & chirurgerie."[11] St. John includes a number of recipes with this statement, including "For the Kings Evel approved" and "For a cough of the Lungues most Approved," and a single use of *probatum* in connection with a remedy to alleviate an ulcer in the throat.[12] In these cases she appears to have used "approved" and such as a general form of endorsement, signaling approval for the recipes at hand. They do not appear to be tied to specific instances of use or singular trials.

In contrast, the verb "to try" appears to have been St. John's choice to record her personal experiences with particular recipes. For example, a recipe titled "For Lime Tree Flower water" comes with this clarification: "I have tryed it sence I had it & raw wort will not doe nither hes smel or tast so that it must be boyld wort."[13] Another recipe to address a "sore that will not heale," obtained from Dr. Richard Lower, comes with this endorsement: "this cured a Leg of mine that had been many months ill in 12 days after I had tryed all ordinary salves & alsoe Cap Greens Powder."[14] Finally, a set of instructions for the "mang" in a dog contained a note that the dosage listed was for a little dog, since Johanna had "tryed" the cure on two of her own animals.

St. John also used the verb "to experience" in relation to recipes that had undergone trials by herself and by trustworthy recipe donors. For example, within her recipe book, a remedy titled "For the Dropsy experienced on my Cooke" suggests that St. John herself oversaw its use and effects. Another recipe for smallpox, given by "Dutchess Hamelton," is endorsed with "& is excellent in all caces in that deseas which she had experience of not only in her own children but very many more beside."[15] Although St. John does not use it in this context, other householders, of course, also used the verb "to experiment." As we saw earlier, Peter Temple titled one of his tables of contents "A Particular Table of proved & Experimented Receits" to highlight those recipes he had personally tried and tested.[16] Temple's use of "experimented" here is similar in meaning to St. John's use of "experience" in that it refers to recipes that had been tested and practiced generally, with no reference to specific, singular events. During this period "experiment" also was at times used synonymously with "receipts" or "recipes," reminding readers of the connection to the medieval meaning of *experimenta*. Within the household recipe ar-

chive, for example, a recipe in Frances Catchmay's book for a rare ointment to "make the face and skin most fayre cleere whyte and smooth"
included an addendum where a gentlewoman acquaintance of the author "tryed another experiment."[17] Much ink has been spilled over the
use and meanings of "experience" and "experiment" in late medieval and
early modern science. Many writers have pointed out that before 1600
the two terms were often used interchangeably. Over the course of the
seventeenth and eighteenth centuries, they came to be distinct, with "experience" referring to information gained from general observations and
"experiment" denoting an actively contrived event.[18] Within the realm of
household recipe knowledge, the meanings historical actors attached to
these two terms remain murky and multiple.

As is evident from my analysis of the St. Johns' recipe book, the terms
householders adopted to describe their recipe endeavors covered a range
of experiential knowledge. Some recipes, particularly those endorsed with
terms like approved, proved, and *probatum est*, were recognized as valuable cures, recommended by trustworthy donors but not necessarily supported by lived experience. While deeming such endorsement worthy of
record, householders also sought more specific confirmation of successful cures supported by brief testimonials garnered from the writer's own
general observations or from experiences recounted by trusted sources.
Reading St. John's differentiated use of terms such as "tryed" and approved, we are enticed to untangle the different categories of experiential
knowledge; however, it is clear that the use of these terms was subject to
the idiosyncrasies of individual writers, so that they often overlapped in
meaning and cannot be neatly boxed up. The abundance of evidence, from
marginal signs to these short statements, attests to the importance of experiential knowledge of various kinds in making recipe knowledge.

Testing for What?

The myriad terms householders used referred to a similarly wide range of
testing practices. Returning to Johanna St. John's use of "tryed," for example, in the three instances listed above, "tryed" pointed to three distinct
uses of recipe trials. In the first case—the lime flower water recipe—St.
John's trials were geared toward assessing the effects of heat on particular
substances. By advising future makers that they needed to boil the wort,
St. John clearly focuses on production. In the second case—Dr. Lower's

recipe for the sore—the trials were comparative, since she recommends his medicine over "ordinary salves" and "Cap Greens Powder." Here the efficacy of the medicine is at the center of the testing. In the final veterinary recipe it was the dosage that was on trial, since she noted, "this is for a little Dog I tryed it to two of mine."

Consequently, householders used recipe trials for a range of reasons. On one hand, the trials appeared to be about refinement, modification, and adjustment—fine-tuning widely circulated recipes to suit individual constitutions and medical practices.[19] On the other hand, they were also ways householders experimented with production methods, equipment, materials, dosage, and more, and explored how these worked in particular environments or toward their immediate needs and goals. Of course, the trials were also about assessing efficacy; as we saw with Conway and Harley's letters, only by making the recipe could one gauge whether it might be of use. A range of processes concurrently underwent inspection in each trial. In the sections that follow, by analyzing interlineal and marginal interjections I detail the various kinds of observations householders conducted during their recipe trials.

One of the most common goals of recipe trials was to test for efficacy. For example, St. John's endorsement of Dr. Lower's recipe for a sore, with the note that, after she had been trying various medicines for months, Lower's remedy cured her leg in twelve days, is a record of both successful and unsuccessful cures. Edward Dering often noted when a particular remedy performed well. Aside from the instances described in this chapter, Dering's notebook is filled with records of successful or unsuccessful cures. In July 1660 he noted that a recipe for a sore leg "did a very great cure on Old Bulbankes legg which the flesh was all gone to the bone and allmost spoiled with old corrupt humours."[20] On another occasion he praised a "country cordiall" as a medicine he had found by "constant experience" to be "of very good use."[21] This kind of practice—trying a cure on a human body to observe its effects on the sick—was common among all kinds of medical practitioners across Europe, from learned physicians to itinerant doctors to noblewomen.[22] And household recipe books are filled with such endorsements of particular cures.

While householders were eager to find "approved" cures and to note down their own, their friends', and their family members' successful experiences with particular remedies, they were also keenly aware of the difficulties of both repeating and replicating successful cures and the very real chance that cures approved and endorsed by others would not work on the case immediately at hand.[23] Several factors determined the "success" of a cure, and a remedy that worked on one person's body in one particular instance might not work on another person or on another occasion. The humoral theory, after all, emphasized the unique balance of humors and temperaments in individual bodies. This kind of "uncertainty" connected to medicine was widely recognized across knowledge communities in late medieval and early modern Europe.[24] With this in mind, householders were open to trying numerous "approved" remedies as they attempted to alleviate sickness and pain. A case in point is the Derings' trying various ointments on their daughter Betty's burned eyebrow and forehead. As the quotation above makes plain, they continually monitored the effects the medicines had on Betty's body and readjusted the treatment accordingly.

In addition to observing a particular medicament's effects on an individual human body, householders also continually tried out particular production methods. A recipe to make white mead in the late seventeenth-century recipe book of Anne Glyd offers a good example.[25] The recipe is rather simple. Makers are told to dissolve six quarts of honey in six gallons of water and heat the mixture over a fire. Once it is about to boil, they should first "scrum" it very well and then add an ounce of mace, ginger, and half an ounce of nutmeg and, if desired, a little sweet marjoram, thyme, and sweet briar. The resulting mixture of honey and spices should then be boiled, cooled, and put into a storage vessel. On reading and making this recipe, however, Glyd had other ideas and decided to modify it by adding yeast. Adding yeast to brewed drinks such as mead, ale, and beer is a fraught step in production, since the temperature of the liquid is key:[26] too hot and the yeast dies; too cold and fermentation will not start. Glyd was clearly aware of this, since she wrote in the margin: "the yest must be put in before it is quiet cold[,] when it is white at once put it in the vessell."[27] This short statement reveals Glyd's own lived experiences with the recipe and her struggles to encapsulate in writing a process that really requires personal experience and observation, or "gestural knowledge," to use Heinz Otto Sibum's term. After all, she could warn future users that they needed to add the yeast when the mixture is not quite cold and white, but how would they know the right shade of white?

In testing recipe knowledge, early modern householders also paid attention to the ways materials interacted or how they might behave in different situations. Anne Glyd recorded a number of such observations in her recipe book. One example is a recipe for "An Excellent black Salve"

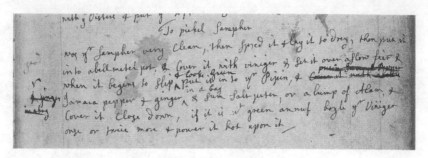

FIGURE 4.1. Wellcome Library, London, MS Western 3082, fol. 68r.

where Glyd subtly changed the production methods. The recipe required makers to heat salad oil, white wine vinegar, white lead, red lead, and beeswax in a pipkin and to boil the mixture until it became "as black as pith."[28] Glyd crossed out these instructions and provided alternative steps in the margin. In her version, makers were told to boil the mixture on the fire "softly" but to "never beate[,] stiring them till it becomes as black as pith." The warning to "never beate" indicates not only that Glyd had made this particular medicine but that she had observed how the materials melted together. Clearly, stirring or beating with too much vigor had been disastrous in her trials of the recipe.

Another instance of how householders might have observed the materiality of ingredients is a recipe "to pickel sampher" in the Johnson family book (fig. 4.1).[29] The recipe, which has been marked as "good" and has a check mark in the margin, called for mixing cleaned samphire with vinegar and heating it until the mixture begins to "slip." Once that happens, makers should put the mixture into a pipkin, cover it with a little Jamaica pepper, ginger, and some saltpeter or alum, and "close it down." In this particular recipe, producing a green mixture was crucial. The original recipe advised makers that if the mixture was not green enough they could boil their vinegar once or twice, then pour it hot onto the samphire. The proper shade of green, however, was left to the makers' discretion.

This recipe, like many others in the Johnson family book, has been worked on and modified by subsequent makers, readers, and writers who recorded their advice and hints in the margins. Here the user warned future makers to watch out for a color change as the samphire/vinegar mixture heated up. As it turns out, the pronounced green color produced was the signal for transferring the mixture to the pipkin for the next step. Additionally, this user advised subsequent makers to put the pepper and

ginger in a bag, presumably to make it easier to separate the final product—
pickled samphire—from the spices used to flavor it. Both these emenda-
tions reveal a writer with firsthand experience in making this recipe. Only
after performing the various steps could a user advise future makers to
take the color change as a signal to move on to the next step and to avoid
the mess of separating the different elements at the end of the dish. Here
the recipe trials allowed the maker not only to ensure that the dish was
"good" and "approved" (which were also written in the margins) but to refine, modify, and improve production methods and to observe how materials interact under particular conditions.

Like many recipe users before and after them, early modern householders made substitutions. Within contemporary medical thought, substitution of ingredients had a long history and even inspired the genre
of "quid pro quo" medical texts—tables offering advice on substituting ingredients.[30] It is thus not surprising that this practice was common
among householders. Edward and Mary Dering, for example, decided to
omit gromwell seeds and spleenwort leaves from a medicine prescribed
by a Dr. Bell of London. Anne Glyd also continually fiddled with ingredients as she tried out the recipes. Next to the instructions for making an
ointment for a new or old swelling, she added the herbs feverfew, savine,
century, and fumitory to the existing list of English and Roman wormwood, peach leaves, southernwood, unset leeks, lavender cotton, tansy,
and smallage.[31] In a recipe for the "imperial water" she noted, "some still
it in halfe brandy and the halfe ale; some put one pint of canary to four
pints of red wine and so I do." According to her marginal note, Glyd also
added two ounces of London treacle.[32]

Even when recipe makers were content with the ingredients listed,
they still played with the amounts and proportions. In the Bertie/Wid-
drington collection, discussed in chapter 5, for example, is a recipe titled
"A purging Cordiall Tinkture of My Lady Sophia Choworths." The recipe
advised makers to steep two ounces of senna and one and a half ounces of
aniseed, coriander seeds, licorice, elecampane roots, guaiacum wood, and
twelve ounces of raisins in *aqua vita* for four days and then strain the resulting mixture. The consumer was told to take two spoonfuls at night and
four in the morning. Obviously not content with the proportions, the same
person who wrote down the original recipe added this note in the margin: "I take only one ounce and a halfe of Sena when the stalkes is piced
out and one ounce of Elicampany root dried."[33] Anne Glyd was also in
the habit of adjusting the proportions and amounts as she made and used

recipes. In a recipe against the "pastion of the heart," she increased the amount of claret used from two spoonfuls to a quarter of a pint and the saffron from a "little" to two pennyworth.[34] Modifications like these were also made to culinary recipes. Within the Fanshawe manuscript, in what was purported to be an excellent recipe for a cake from Lady Butler, the amount of flour used was changed from three pints to three quarts while all the other ingredients remained unchanged. This, one can guess, significantly changed the texture of the cake, and most likely the viability of the recipe.[35] Perhaps predictably, this recipe was later crossed out.

Householders did not stop at trying out new production methods, ingredient combinations, and proportions but also played with the dosage and application methods. Here again Anne Glyd provides us with several examples. In a remedy for pleurisy, the original recipe suggested the patient take one spoonful of linseed oil every two hours for three cycles; Glyd, however, specified that male patients should take two spoonfuls.[36] In making these changes she showed an awareness that drug dosages needed to be adjusted to different bodies—old or young, male or female. In this instance it is unclear whether Glyd made these adjustments from theoretical calculations or had actually seen the medicine used with a patient and realized that a higher dosage was needed.

Aside from changing dosages, householders also fine-tuned the ways certain medicines were applied to the body. Peter Temple, whom we met in chapter 3, provides a rare example of how householders might have refined the methods of applying medicine. As we learned, Temple had three recipe books—one *vade mecum* in which he entered recipes as he received them; an intermediary book into which he copied recipes he was testing; and a "neat" presentation book that he copied out as a gift for his daughter Elinor. The three books contain a number of overlapping recipes, and comparing them shows Temple's testing practices. The recipe of interest here is one for Lady Tyrell's eye water, which exists in three versions across the three notebooks. Comparing these shows Temple's close engagement with the recipe over multiple occasions as he strove to improve the method of applying the medicine.

We first encounter the recipe—short and roughly scribbled by an unknown hand—in Temple's portable *vade mecum* (fig. 4.2). Although no author attribution was written alongside the recipe itself, a list titled "Chuse receits are & approved" at the beginning of the notebook contains an entry for an eye water by a "Ja. Bynnetton."[37] Like many others in this chapter, the recipe is simple. The method involved putting two drams

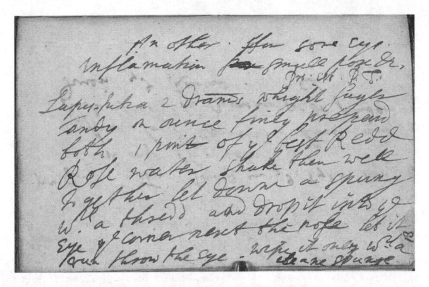

FIGURE 4.2. © The British Library Board, MS Stowe 1079, fol. 81r.

of lapis tutia and one ounce of white sugar in one pint of red rose water, shaking well, and applying the mixture to the eye. In this version of the recipe, only brief application instructions were provided: "drop it into the eye the corner next the nose let it run throw the eye. Wipe it only with a cleane spunge."[38]

A second version of the recipe resurfaces in the intermediary book, written in Peter Temple's distinctive hand (fig. 4.3). Here the scrappy notes from the *vade mecum* have been transformed into elegant prose and the recipe is attributed to "Lady Elinor Tyrrill fr[om] James Bynnetton. It hath commendable effects and without payne." Additionally, readers were told that "in case of extremity" they might consider substituting "alloes" for the sugar. The instructions on how to apply the eye water were also far clearer. First, users were told to tie a piece of thread around a sponge and drop the sponge into the bottle. They should then shake the mixture well and "drop it into the patients Eye at the corner next his nose and lay his head soe if it may run throw his Eye—let him wipe it with the like spunge dipt in the same water but wipe it toward the nose to avoid blearing wash the spunge as oft as you use it."[39] The careful and detailed instructions to use clean sponges in order to avoid blearing—watery or inflamed eyes—indicates an awareness of possible contamination or further irritation of the eye. In this version the practitioner

FIGURE 4.3. © The British Library Board, MS Stowe 1078, fol. 25r.

was also advised that laying the patient flat would make applying the eye drops easier.

The final version of the recipe in the "gift" book sees Temple further refining his application technique and codifying it (fig. 4.4). In this version readers were told to "take a little peece of a spung very cleane, tye a thread about it, then let it into the glasse and shake these ingredients well together as oft as you use it." Then they should "cause the patient to lye on his back with his head on a pillow." Using the sponge, they now "may dropp the water into that corner of the eye next the nose, and soe let it passe cleane throw the eye (or else it is ineffectuall)." Additionally, they were told to use a clean sponge every time they wipe the eye, taking care to begin from the "corner farthest from the nose, and soe draw it inward least you blear the eye."[40]

Besides expanding and clarifying the application instructions, in the transfer from the intermediary book to the gift book Temple also made other substantial changes to the text. First, as is evident in figure 4.4, he encoded the name of the recipe donor and the ingredients into a simple cipher. As described in the previous chapter, he used this code on most of the recipes in the gift book to ensure that the knowledge it contains would not fall into the wrong hands. Second, in contrast to versions of the medicine in the *vade mecum* and the intermediary book, the medicine was now assigned solely to Lady Tyrell. It is worth noting here that since Peter's wife was from the Tyrell family, the Tyrells were frequent donors to the book and, one presumes, significant relatives of the Temples. Third, the substitution of "alloes" has been dropped, and we get a further qualifier for the red rose water: "the buds is best."

This eye water recipe is significant not only because Temple prized it but also because the three versions clearly demonstrate the hands-on practice needed in *using* remedies. Many of these rather minute changes to the application method may seem trivial, but the continual rewriting of

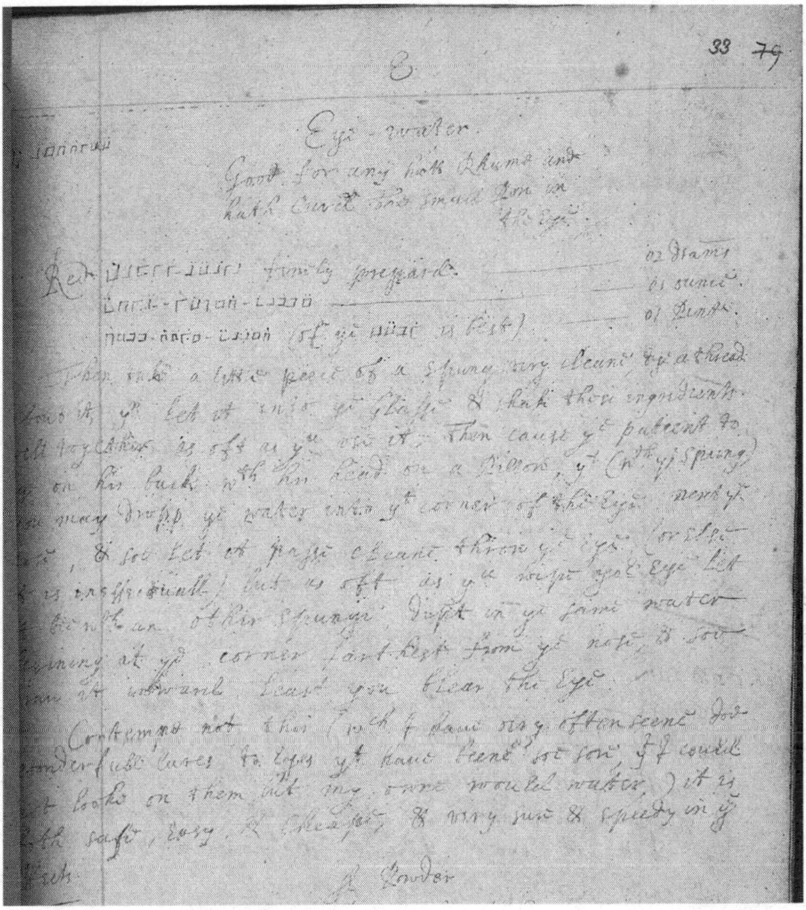

FIGURE 4.4. © The British Library Board, MS Stowe 1077, fol. 33r.

these instructions shows a writer who clearly has tried applying this eye water to sufferers. Temple's own observation and experience with the cure is further emphasized by his endorsement: "I have very often seene doe wonderful cures." In thinking about how recipe knowledge was created, we often focus on the know-how required to produce the substance—the maker's familiarity with distillation processes or the brewer's expert skill in knowing exactly when to add the yeast. In this example, however, we get a sense that even with the simplest medicines (in this case the production method is "shake well"), a great deal of tacit knowledge was required for applying the cure. The continual refinement of the method suggests that in this case Temple's trials were directed to this particular area.

The many examples listed in this section demonstrate how early modern householders, through making and using medicines and foods, explored written recipe knowledge in various ways. Encoded in the margins and between lines of texts and recipes are a series of rich practices. Some were part of the householders' schemes to customize medicines to suit individual bodies or create foods to cater to their own tastes and palates. Through other practices recipe makers gained a deeper understanding of how materials behaved in particular circumstances—of how they interacted with each other and reacted to external stimuli. As such, their investigations are part of a wider interest in materials and materiality across Europe during this period.[41] Finally, as with Peter Temple and the eye water, these practices let them gain, through doing, the skills and knowledge needed to apply medicines to the body. Much the way Anne Glyd's white mead recipe hints at the embodied knowledge required to brew ale, beer, or mead, Temple's eye water recipe reminds us of the lived experience required to use the simplest medicines.

While recipe writers are fairly vocal about whether a recipe is "good," the broad sphere of home-based recipe trials and tests indicates that householders were rather open-minded about assessing individual recipes. In most cases the trials and assessments worked concurrently in a number of areas, from production methods to ingredients to effects on the human body. After all, recipe makers could mark a recipe as "good" yet suggest myriad changes. What might be considered a failure when producing one substance might later be reviewed as a successful way to make another substance. For example, in the recipe book associated with Elizabeth Godfrey that we encountered in chapter 3, a recipe for syrup of clove gillyflower has been adapted by later users to make syrup of damask roses (fig. 4.5).

The recipe was repeatedly scrawled over on different occasions. In the title, one annotator cleverly rewrote "clove" to "Rose" and crossed out "gilly flower" entirely. The method was then adapted to make this new medicine. The adaption required only changing the main ingredient (from clove gillyflower to rose), the ingredient proportions, and the equipment used (from silver tankard to deep gally pot).[42] The main method of the recipe, though, remains intact. Evidently, even if the clove gillyflower syrup was ineffective or unwanted, users were still keen to retain the technique. Examples like this gillyflower or rose syrup demonstrate that early

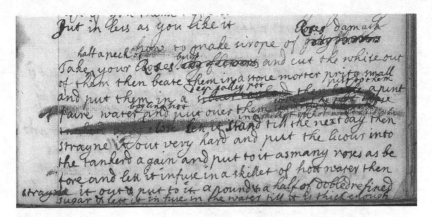

FIGURE 4.5. Wellcome Library, London, MS Western 2535, fol. 81.

modern householders not only took a multipronged approach to their recipe trials but also kept a certain flexibility in the assessment criteria as they moved from recipe to recipe and from trial to trial. Here the medicine might not have passed the test, but the method certainly gained approval. Undoubtedly this is in part due to inherent uncertainties in recipe knowledge. As I noted above, the messiness of the human body has always privileged personal experience in establishing the efficacy of particular remedies on individuals, so that ascertaining whether these properties might transfer to other bodies was fraught.[43] Likewise, as scholars have pointed out, premodern materials, equipment, and instruments each brought new issues and required recipe makers to work through instructions and adapt them to the limits of their own workshops.[44] Added to this, as has been shown in earlier chapters, household recipe knowledge was framed by contemporary notions of sociability and economies of regard and obligation. As we will see, the social value of recipes and the social function of recipe exchange shaped householders' recipe trials and prompted some to recipe salvage.

Recipe Salvage

There are a number of instances within the recipe archive where we see the readers/makers repeatedly try and tinker with a recipe only to reject it at the final stage. An example is Elizabeth Godfrey's recipe for making

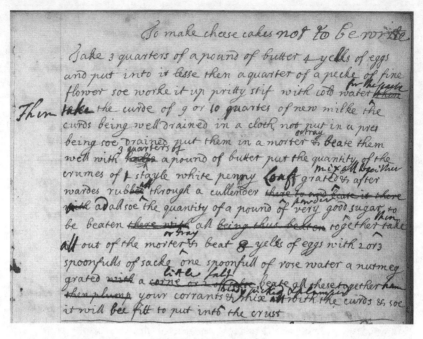

FIGURE 4.6. Wellcome Library, London, MS Western 2535, fol. 5.

cheesecakes using the curds of new milk[45] (fig. 4.6). As is evident in the image, the recipe has been worked over by two annotators. The first, perhaps the same person who wrote the recipe down, increased the amount of butter needed (from half a pound to three-quarters), then advised future makers to beat the curds and the butter together in a tray. Subsequently a second reader/annotator/maker also worked through the recipe. The second annotator reduced the egg yolks required from twelve to eight, then streamlined the recipe by advising makers to first mix the butter, curds, and breadcrumbs and then rub the resulting mixture through a colander. This differs from the original method, which called for grating a stale white penny loaf and pressing it through the colander. The method the second annotator suggested would no doubt make a much smoother paste. Not content with this, the second annotator also reminds makers that the currants for the cakes should "be ready picked and plumped."

The changes to this cheesecake recipe are multifold. Together the two annotators suggested changing equipment (from a mortar to a tray—

perhaps the curds of nine or ten quarts of new milk would not fit in the household's existing mortar), adjusting the ratio of cheese curds to butter, reducing the egg yolks required, *and* changing the production method (pressing everything through the colander for a smoother texture). These two annotators not only have actually made the recipe but also have actively engaged with the know-how to refine and improve it. Yet even after all these changes the recipe has been labeled "not to be write." Recall from chapter 3 that Godfrey is one of the recipe compilers who worked with multiple books. Consequently, her "not to be write" suggests that the recipe was not to be copied into the family's trusted book of recipes. Ultimately, despite all these changes, the annotators must have decided it was not salvageable and thus declined to incorporate it into the family treasury of knowledge. Curiously, though, before it was deemed unfit, they took considerable time to try to improve the recipe. This know-how was certainly not rejected on a whim.

Another example of recipe salvage is offered by the readers/makers of Anne Glyd's book. The recipe at hand is titled "An exselent ointment for any hott inflammation." It is covered by writing in four distinct shades of ink, suggesting that the readers/makers of this recipe collection repeatedly returned to this set of instructions, adding to it and modifying the know-how. This recipe has also been crossed out, signaling that ultimately it too was deemed unsalvageable.

The recipe as originally written was fairly simple. Readers were told to shred six different herbs and boil them in two pounds of unsalted butter or fresh grease.[46] The annotators made several changes. First, right under the title, an additional note told future readers/makers that if mixed with a little honey this ointment would also be good for the ague and St. Anthony's fire. Put in the mouth, it alleviated toothache. The second change is in the production method. Here readers were told that instead of merely boiling the herbs in the fat, they should boil them a little at night, let the mixture stand overnight, then boil and stir it the next morning. Finally the recipe, initially just signed as "probatum Ann Glyd," was now also attributed to "aunt eavens." As with the other examples in this section, we see modifications on several levels, from the use of the recipe to the production methods to the attribution. Alas, after all this work the recipe was still rejected with a rather forceful double cross-out.

The Johnson family book provides us with a final example of such attempts at recipe salvage (fig. 4.7).[47] In this instance the recipe offered instructions for making gooseberry vinegar. The instructions, before and

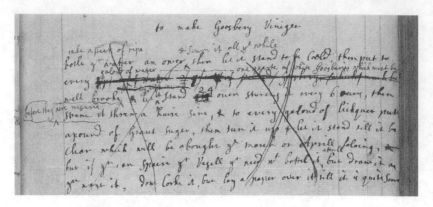

FIGURE 4.7. Wellcome Library, London, MS Western 3082, fol. 65r.

even after the modifications, are rather confusing. The original recipe required makers to boil water and cool it, then add "well brooke" (one assumes that means crushed) gooseberries, leave the mixture to stand for twenty-four hours (stirring every six hours), strain it, add sugar, and then bottle it (closing it not with a cork but just with a piece of paper) until the vinegar is "quite soure." The recipe, like the cheesecake recipe described above, was scribbled over by a later reader/user and subsequently crossed out entirely. Reading the annotations, it is clear that the later reader/user made valiant attempts to salvage the recipe, changing the proportion of water to gooseberries and inserting extra steps to skim the boiling water and rest the mixture before adding the sugar. The reminder to skim ("scum") the water suggests that the annotator noticed that water boiled on its own would form a scum that had to be removed, and altering the water/gooseberry proportions indicates that the recipe might have undergone a taste test, leading the householder to decide on a slightly different way forward. Thus both these changes show that this particular reader/user had experience with making gooseberry vinegar from this particular recipe.

The Godfrey and Glyd examples were instances when, even after extensive modification, makers and writers deemed a recipe unsalvageable. In contrast, after much modification the Johnsons arrived at a version of the gooseberry vinegar recipe that they were happy with. On the verso of the same folio is a recipe titled "to make Sister Jacksen Goosberi Viniger which is very good."[48] The recipe is twice marked with "good" and a check mark in the margin and offers a variation of the crossed-out instructions described above. The basic production process remained the same—makers need to boil and cool the water, add crushed gooseberries, let the mixture stand for twenty-four hours (stirring now and again), add sugar, then pour the mixture into a bottle, using a paper stopper. However, the Johnson family continued to modify individual steps, and this second recipe is also scrawled over with annotations and changes. Most helpfully, a "nota bene" explained that "a peck of gosberries makes very near 4 quarts when they are brused, so that quantity requires near 4 ga-lands of water, by which ye may know how to make the quantity, ye chose to make." There are no dates or time stamps in the recipe book, but the various inks affirm that the Johnsons made repeated trials over a signifi-cant period, testing and refining the know-how. Ultimately the family's efforts at recipe salvage were rewarded, since they finally created a recipe they could describe as "very good."

These three examples highlight the practice of salvaging recipes. It is clear that once a recipe passed the stage of trial on paper described in chapter 3, householders like the readers/users of the Godfrey, Glyd, and Johnson manuscripts were willing to invest a great deal of time, resources, and manpower to test, modify, and retest instructions. A sunny optimism permeates such practices, with a belief that if one made enough changes, any recipe could be salvaged: a recipe unsuitable for producing syrup of clove gillyflowers could be adapted to make syrup of damask roses. Within this setting, recipes were viewed as open sets of instructions, and what might count as success or failure was malleable. The testing of know-how within the household was tightly bound up with contemporary prac-tices of modifying and customizing. Adapting the instructions to make a different kind of flower-based syrup might also reflect changes in fashion and personal taste. It may be that later users of the recipe book simply preferred rose-flavored syrup, for its taste or for medical reasons, over one flavored by gillyflower.

Another reason for this recipe salvage might be traced back to the social dimension of recipe collecting and exchange described in chapter 1. Using recipes in gift exchanges and in building alliances and networks had both an epistemic and a social value. Recipe collectors therefore judged the po-tential of recipes by a range of criteria, and medical expertise was only one consideration. Within this view, testing a particular recipe was also testing the trustworthiness of a particular knowledge donor, not only in medical matters but also more generally. For example, Peter Temple noted next to a remedy to bring away the afterbirth, "Sir Alexander Hambleton . . .

saith the Author whome I have found a man of trueth in other matters."[49] Within this model of knowledge making, compilers might have been encouraged to modify and salvage recipes because of the high social value afforded to particular sets of instructions; they may have wanted to maintain and uphold social relationships even though the donors' know-how was less than ideal. After all, it was not only medical expertise and authority at stake but at times much more. It is no wonder householders tried and tried again before dismissing sets of know-how. When Anne Glyd crossed out the ointment recipe, she was not only rejecting know-how but also questioning her Aunt Eavens's authority and expertise. Considering that the gooseberry vinegar recipe was donated by "Sister Jacksen," the Johnson family's relationship with her might have increased their efforts to salvage the recipe. In a way these concerns also framed the assessment of recipes at the stage of "trial on paper." A hesitancy to taint his relationship with Harley and with Sir John Tracy might partly explain why Conway went to such lengths to investigate, explore, and examine the ale recipe at the center of chapter 3.

Rewriting Recipes, Making New Recipe Knowledge

Early modern householders' continually making, testing, and modifying recipe knowledge had other far-reaching consequences. Hitherto we have encountered them modifying, approving, or rejecting recipes in the same location—on the same page in a notebook. This paperwork enabled users to retain the different layers of modification and so the ability to revert to earlier versions of the recipe. As is evident from the analysis presented in this chapter, the original recipe, in spite of the cross-outs and the multiple annotations, remains visible, consultable, and usable. Yet this multilayered approach was not the only model for writing down recipe knowledge. In some instances recipe writers might choose to incorporate modifications, additions, and suggestions into the main recipe and then write a new version. With this approach traces of past tests, trials, and, crucially, recipe authors are erased as new recipe knowledge is made. Peter Temple's three notebooks contain multiple instances of this model. This section will explore two examples: Sir Roger Nicholls's recipe for a purging ale and Sir Alexander Hamilton's recipe for a balsam. Both these recipes held central places in Temple's medical practice. They were copied across the different notebooks, included in the list of "choyse receipts" written out in the *vade mecum*, and recommended as key medicines to have on hand.[50] These recipes are ones Temple was very familiar with and that he had made, used, and tried over and over again.

Sir Roger Nicholls's purging ale is a diet drink purported to "Clense the Body of all Corrupt Humours, water & wind," make "the Countenance Cheerefull & the body lightsome," and more. The recipe involves steeping a number of purging materia medica (including sena, rhubarb, aniseeds, sassafras wood, and more) in ale for three or four days, after which the ale is mixed with more herbs (betony, maidenhair, brooklime, watercress, and agrimony) and the juice of scurvy grass. The resulting drink was to be taken twice a day for about a month every spring and fall. Two versions were recorded in Temple's notebooks—one in the intermediary book and another in the gift book. A number of changes were made during the rewriting or copying of the recipe. First, Temple gave it a new title. Whereas the recipe in the intermediary book is called "An approved Medicine called the Purging Ale," in the gift book it is titled "A Dyet Drink," with the following qualifier: "I have oft approved successfully on severall occasions heereafter mentiond with my selfe & Friends."[51] Sir Roger Nicholls is still acknowledged in the top left corner of the page, but with this qualifier: the authority of the recipe in the gift book more properly rests with Temple than with Nicholls. The other change sees information initially recorded in the margin shifted into the main text. In the intermediary book Temple noted in the margin, "we dranke a pint in the morning and halfe a pint at night P.T., E. T." and "Mr Nicholls spoke since of Harts tongue to be added." Both these marginal annotations reflect contemporary recipe practices. The first not only serves as a record of Temple's own experiment with the cure but offers personalized dosage information. The second reminds us that recipe exchange was framed within continuing personal relationships and conversations about recipe knowledge. As we saw in chapter 3, Temple's notebooks are filled with reminders to inquire further about particular recipes. This marginal note offers a rare glimpse of the multiple interactions early modern householders might have had over recipes. Both the changes were incorporated into the main text in the gift book; thus, during the final step in the codifying, modifications once regarded as marginal were fully integrated into the recipe.[52]

Our second example, Sir Alexander Hamilton's recipe for a "wonderous balsome," again sees Temple "reattributing" recipes in his multistep system of codifying knowledge. The recipe can be found in all three Temple recipe books. The versions in the *vade mecum* and the intermediary book are both attributed to Sir Alexander Hamilton and show only minor discrepancies.[53] In the gift book the same recipe is labeled "*My* Balsome for wounds etc." (my emphasis). Here the recipe is accompanied by an account of Temple's discovering the know-how in France and is described as "The most universall medicine in this book."[54] Along with the change in authorship came a number of changes to the text. The gift book version has a number of additional ingredients, including thyme, balm, cochineal, Spanish wine, ambergris, mastic, myrrh, guaiacum, and sarsaparilla. The final two ingredients, Temple specified, "must make it good for the French dissease."[55]

Perhaps the most significant changes to the recipes were in the list of the medicine's "virtues." From the start, the medicine was a universal panacea, with more than twenty uses. In the gift book version the list of uses was hugely expanded to forty-two. Originally the cure addressed a range of ailments from aches and strains to sores, ulcers, and wounds to burns to the gout to sciatica to stomach pains and headaches. The additional uses covered an equally wide range of afflictions, including measles, plague, fistula, sore breasts, dropsy, cold, ague, consumption, bloodshot or sore eyes, and much more. Significantly, a number of these newfound uses were accompanied by accounts of successful cures. Some referenced the balsam's curative powers on Temple's own body—"It cured me of a Bloody-flux (after divers medicines had fayled for 3 weeks together)"—and others on someone else's body—"It is good for old itches . . . (I cured one of 40 years standing P. T.)."[56] While these two examples reference Temple's personal observation of the cure, other accounts are more vague. For example, virtue number 26 claims the balsam was "good for a Consumption by eating a Pill as bigg as a nutmegg morning and evening roled in sugar. The La: Crumpton was cur'd when the D^{rs} gave her over, and when she spett nothing but blood, and tis better if you take the same quantity of conserve of Roses with it."[57] By offering descriptions of his own or other trustworthy witnesses' experiences as proof of efficacy, Temple roots his recipe knowledge making in practices of testing and trying. With the support of these trials, the balsam, once touted as Sir Alexander Hamilton's wondrous cure, was fully transformed into Peter Temple's own formula. This kind of authorship reassignment was common across other writing genres in the period and reflects contemporary attitudes toward texts as well as the inherent malleable quality of recipes as text.[58]

Conclusion

Edward Dering's recipe-filled summer and the numerous examples outlined in the previous two chapters suggest that early modern English householders were constantly busy conducting recipe trials. Sometimes these trials were driven by mere curiosity (Dering's "experiment" of distilling stale mead comes to mind), and at other times they were brought on by an immediate need such as Betty Dering's burns. Early modern household recipe collections are complex written records of such home-based knowledge practices. Recipe collectors searched widely to gather recipes, but they also tested and tried them before incorporating them into their own canon of useful knowledge. By attending to marginal marks and interlinear additions, this chapter has highlighted the rich set of hands-on practices used to assess and examine recipe knowledge. Three major themes emerge from the margins and between the lines.

First, while trying and testing recipes was clearly widespread and important among householders, the aims and goals of these trials were wide-ranging. While testing for efficacy was clearly an important strand, the home-based trials involved more than just observing the effects of a medicament on the human body. Rather, they entailed testing and continually modifying methods, ingredients, dosages, and application procedures. The range of modifications includes substituting ingredients or changing their proportions according to personal taste or product availability and also modifying production methods for reasons either practical (lack of specific equipment) or theoretical (Conway's objection to boiling plain water for three hours). This is in part because what might be considered efficacious and successful, or a "good" recipe, depended on so many factors—how the medicament reacted with a particular body, which ingredients particular households preferred, whether the householders could follow a particular set of instructions with the equipment and ingredients on hand, and more. The goals of testing and trying thus varied from household to household and from recipe to recipe.

Second, as the section on recipe salvage suggests, the social and cultural functions taken on by recipe exchange and household recipe books colored the way our actors tested and tried medicines. Adopting or rejecting a piece of know-how was inseparable from maintaining one's personal relationship with the recipe donor. In addition to the openness with which householders viewed issues of efficacy and success, this meant that in some cases many were content to continually try out recipe knowledge.

After all, as we recall from the Godfrey family's repurposing the gilly-flower syrup recipe to make damask rose syrup, recipes could be deemed a success even if one adapted the instructions to make an entirely different product.

Third, in emphasizing hands-on trials and personal observations of know-how, householders were participating in European-wide practices conducted across a number of settings. When it comes to medicine, this kind of testing based on evidence gathered from sensory experience has long played a key role. Late medieval physicians such as Arnold of Villanova and Bernard of Gordon composed sophisticated and complex rules for testing drugs and trying cures, with the senses—taste, smell, and sight—at their heart.[59] Drug trials based on gathering experiential knowledge continued throughout the early modern period, and practitioners of all stripes, from physicians to apothecaries to itinerant doctors, conducted a range of tests to address many of the same questions preoccupying the householders featured here.[60]

Concurrently, craftsmen and artisans were also continually testing, modifying, and assessing practical know-how. Making and knowing, as Pamela Smith has argued, went hand in hand in this period. The focus of this book on the early modern home and on collaboration among families and household members extends our current narrative into new spaces, with a new cast of characters. For historical actors such as Peter Temple, Edward Dering, and Anne Glyd, testing, trying, and approving recipes was part and parcel of household science. As was true for artisans and craftsmen, their recipe trials were a way to further their understanding of natural materials and technical processes.

While personal experience and direct observation stand at the center of the practices described here, our historical actors are reticent about contextualizing their undertakings within theoretical frameworks. Even in the letters exchanged between Edward Conway and Edward Harley, where the multistep assessment system and the reliance on experience and trial is made clear, there are few hints of how these trials might fit into any epistemological scheme. In this sense, even though these practices were occurring at the same time as the natural philosophical experiments conducted by members of the nascent Royal Society, they had different aims.[61] Yet while it has received a great deal of scholarly attention, experimental philosophy, as Ursula Klein and Wolfgang Lefèvre have pointed out, was only one of a number of interconnected experimental styles in the premodern period. Alongside experimental philosophy, technological inquiry and experimental histories—or *historia experimentalis*, as detailed by figures such as Francis Bacon or Robert Boyle—constituted important and interconnected pathways for hands-on investigations of the natural world. These schemes "collected, described, and ordered facts relating to the perceptible dimension of particular objects and processes" and "reported phenomena procured by intervention into nature, both in arts and crafts and academic sites."[62] For practitioners of experimental history, then, the goal was less to formulate conceptual frameworks or causes than to collect experiments, observations, and descriptions of natural phenomena. In their focus on materiality and production methods, the everyday recipe trials of householders have much affinity with the goals and practices of early modern experimental histories, particularly those in early chemistry, where much the experimenting took place in kitchens and stillrooms. Consequently we might read the householders' eager and diligent gathering of natural knowledge from recipe trials alongside these schemes. Together they constitute an epistemic culture that sought to piece together, recipe by recipe or fact by fact, a richly textured picture of the natural world.

CHAPTER FIVE

Writing the Family Archive

Recipes and the Paperwork of Kinship

In 1931, led by Phyllis Brockman, a group of men and women pried open two enormous chests in the country house of Beachborough Park in Newington, Kent. Peering into the chests, which probably had not been opened in the lifetime of any of them, the group found bundles of land deeds, records of marriage settlements, transcripts of lawsuits, account books, diaries, travel journals, notebooks filled with exercises in arithmetic and French grammar, game books, and hundreds and hundreds of medical, culinary, and other household recipes.[1] The papers belonged to the Brockman family, who had made Beachborough their home for more than two hundred years before selling the property in the eighteenth century.[2] Later that year Phyllis donated to the British Museum more than 2,500 documents dating from the late thirteenth century to the nineteenth.[3] The bulk of the donation consisted of early modern legal and financial papers centered on the lives of Sir William Brockman (1658–1740), his wife, Anne Glyd Brockman (1658–1730), and their son, James. Despite the size of the deposit, this initial gift did not include all the papers from those two enormous chests. Phyllis kept twenty-eight volumes for her own and her family's personal consultation. Among the papers she found so hard to part with were the recipe books of four Brockman women. The earliest book, a leather-bound quarto volume, had belonged to Ann Bunce Brockman (1616–1660), wife of Sir William Brockman (bap. 1595–1654), a royalist who had played a significant role in the Civil War and who shared recipe knowledge with Edward Dering, the protagonist of the previous chapter.[4] A smaller volume is connected with Ann's granddaughter Elizabeth (d. 1678), who filled the book with both culinary recipes and a series of

French exercises.[5] The last two collections were created by Anne Glyd Brockman and her mother, Anne Glyd. They include a bound folio-sized notebook consisting largely of recipes gathered by Anne Glyd and her family, which featured centrally in chapter 4, and a series of about a hundred loose recipes dating from the eighteenth century, many addressed to Anne Glyd Brockman.[6] Taken together, the recipe archive Phyllis Brockman uncovered spans three generations and multiple families.

When Phyllis Brockman and her helpers broke open the great wooden chests in 1931, Beachborough Park had long fallen out of the hands of the Brockman family and was the site of a private preparatory school. Yet everyone concerned agreed that the Brockman family papers belonged to Phyllis and, unlike the physical building of Beachborough Park and its extensive grounds, were still part of her inheritance. By the 1930s, much of the medical knowledge, and indeed culinary knowledge, presented in these volumes was outdated. It is likely that Phyllis's interest was motivated by her desire to learn more about her family rather than to learn how to cure a cold or a cough or to bake a cake. Her other selections from the chests of family papers confirm this view. Alongside the four recipe books, she chose to keep Richard Glyd's testamentary instructions made the year before his death in 1666; Glyd, Stoughton, and Drake family genealogies; James Brockman's university exercise books; William Brockman's journal detailing his travels in Germany and the Low Countries; and a twenty-four-year run of personal accounts from her ancestor Anne Glyd Brockman.[7] This reading list hints at a fascination with her ancestors' private lives, personal pastimes, and family history.

Prompted by the Brockman papers, in this chapter I explore the multitude of ways early modern householders used recipe collections as a way to construct and write their own family histories.[8] Two connected historiographic strands frame this chapter—the study of family strategy and of archival strategy. Scholars of late medieval and early modern Europe have drawn our attention to the many ways landed gentry took firm control of their family fortunes by encouraging particular marriage alliances, designating heirs, and cultivating and maintaining particular social and family networks.[9] Writing down family histories and creating family archives were central to these practices. In early modern England, gentry and other landed families, as Daniel Woolf and Katharine Hodgkin have argued, affirmed their positions within the local communities and national and courtly networks by carefully crafting a coherent narrative spanning generations and by preserving and documenting family wealth in material, social, and financial terms.[10] Concurrently, historians of archives, particularly those focusing on institutional archives, have also written cogently on the notion of archival strategies.[11] Archives, as historian Kathryn Burns has argued, "are less like mirrors than chessboards" created by a series of "gambits, scripted moves and counter moves."[12] Although Burns writes on institutional archives in colonial Cuzco, we may extend this analogue to household archives in early modern England. Gentry families, who are at the center of this book, carefully preserved and curated legal and financial papers to safeguard their family estates and holdings and to fashion a family identity, particularly a genteel identity. The inclusion of recipe collections among the legal and financial records of the Brockman family indicates that recipe writing was firmly a part of wider practices of household archiving. The impulse that led collectors and makers of recipe knowledge to ensure that their diligently gathered and tested know-how was passed to the right person was also shared by men of science. Naturalists such as John Aubrey and John Evelyn were constantly preoccupied with strategies and maneuvers to preserve their intellectual legacies.[13]

Recipe writing, as other scholars have established, was one way writers constructed identities.[14] Here I extend that view and widen it to examine the early modern household as a unit and to situate recipe writing within the creation of household archives. I argue that gathering recipes and creating recipe collections was one aspect of what we might call the "paperwork of kinship." Early modern householders created, collated, and preserved all kinds of paperwork concerning the social and economic holdings of the household, from land deeds to rent accounts to lists of births and deaths. Recipes and recipe books played a crucial role in this paperwork. Working together, these documents not only sketched out a social and economic history of the family but also constructed the family's very identity. This chapter opens with a discussion of the value our historical actors placed on recipe books and links this perceived value to records of family history and social connections. The rest of the chapter introduces the concept of the paperwork of kinship and examines strategies early modern men and women used to inscribe, record, and preserve recipe knowledge and to construct notions of family.

Valued Objects: Keeping Recipe Books in the Family

Although Phyllis Brockman seemed to have inherited her family papers quite by happenstance, surviving evidence suggests that many early modern men and women took a keen interest in who got ownership of such papers. Particularly for recipe books, there is ample evidence that householders strove to control the transfer of the objects, attesting to the value they placed on recipe knowledge. In chapter 1 we saw how Johanna St. John made specific bequests to ensure that her recipe books ended up with the right family members. Another mother, Lady Frances Catchmay, took no chances in ensuring that *all* her children would inherit the practical household knowledge she had gathered over the years. A copy of her collection bears this inscription on the first folio:

> This Booke with the others of medicins, preserves and Cookerye, my lady Catchmay lefte with me to be delivered to her sonne Sir William Catchmay, earnestly desiringe and charginge him to lett every one of his Brothers and Sisters to have true coppyes of the sayd Bookes, or such parte thereof as any of them doth desire. In witnes that this was her request, I have hereunto sett my hand at the delivery of the sayd Bookes. Ed. Bett.[15]

The gift of the "family books" to William Catchmay came with the responsibility to disseminate the knowledge they contained. Frances Catchmay's detailed instructions reveal both the great value the Catchmay family put on recipes and the way sharing the same set (true copies) of practical knowledge helped to maintain ties between siblings.

Copying out collections of recipes as outlined in the Catchmay bequest was fairly common. For example, Charles Howard (1630–1713) commissioned a copy of "Mrs Corlyon's book" for his niece the "Marquesse of Valpariso" in 1679.[16] Householders also made their own copies of their relatives' books. Mary Grosvenor, for example, copied a section out of her grandmother's book and ended her copy "heare endeth the first booke, beinge my granmother Cholmeleys Medisines."[17] Grosvenor's careful notation of the original sources for the recipe section demonstrates her high regard for her grandmother's expertise and for their family relationship.

Getting one's hands on the family recipe book mattered greatly to many early modern men and women. In some cases, such as the Johnson family from Spalding, Lincolnshire, the second generation felt it was their right to inherit the "family book." Maurice Johnson, an antiquarian and barrister, was particularly keen to assert his claim to his stepmother's book of recipes. The volume now has three ownership notes. The first, "Elizabeth Phillipps 1694," was written before Elizabeth Oldfield Philips married into the Johnson family. The second, "Eliz Johnson the gift from her Mother Johnson," was most likely written by Maurice Johnson the younger's wife,

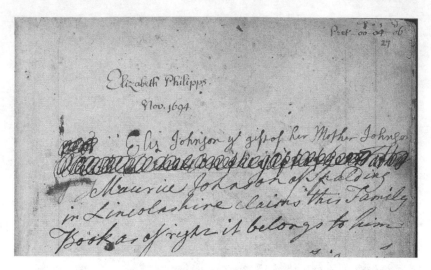

FIGURE 5.1. Wellcome Library, London, MS Western 3082, fol. 27r.

Elizabeth Ambler Johnson. The final note, written by Maurice Johnson the younger himself, reads "Maurice Johnson of Spalding in Lincolnshire claims this Family Book as of right it belongs to him"[18] (fig. 5.1).

Johnson's enthusiasm for household practical knowledge may be related to his other intellectual interests, since he was a founding member of the Spalding Gentlemen's Society of Lincolnshire. The society is one of the oldest such organizations in England, and in its heyday it boasted both a museum and a physic garden.[19] After he claimed the book as his own, Maurice Johnson and his immediate family certainly made good use of it. It became the work of a large number of compilers, who reviewed and annotated particular recipes, and new information was still being added in the mid-nineteenth century.[20] As the fruits of the labor of generations of Johnson family members, Maurice Johnson's terming it the "family book" seems apt.

The many examples cited in this section demonstrate that inheritance and bequest were the main mechanism for transferring recipe knowledge. Much the way learned scholars or natural philosophers might dictate the afterlives of their working papers, the creators and users of recipe books carefully controlled present and future access to their stores of knowledge.[21] In the realm of recipes and recipe knowledge, this had several consequences. As the "family books" were passed from generation to generation, each new owner was encouraged to augment, modify, and customize them; making recipe knowledge was thus collaborative over long periods. In his discussion of the relation between commonplace books and life writing, Adam Smyth argued that since many were created across several generations, commonplace books can be seen as written by a family and considered a depository of a family's wisdom.[22] Recipe books might be thought of along these same lines.

As the Johnson and Catchmay families demonstrate, the inheritance of family books often crossed gender lines. Early modern mothers, fathers, sons, and daughters alike plotted and planned to control the survival and *fortunes* of the family's book of recipes and its store of recipe knowledge. Such consideration was extended not only because the recipe knowledge the notebooks contained was, as illustrated in chapters 3 and 4, gained through firsthand testing and trying and adapted to tastes of the family but also, as outlined below, because recipe books recorded both family history and the social connections and alliances of the household.

Documenting the Work of a Family

One of the four recipe books Phyllis Brockman saved was the book of Anne Glyd, familiar to us from earlier chapters. Although the book is largely in Anne Glyd's own hand, other family members also made significant contributions. Before Richard Glyd died, he was Anne's partner in gathering household knowledge, and he wrote down several recipes he had collected or devised, including two remedies for horse ailments.[23] Recalling the gendered division of labor in the St. John family, where it was Sir Walter who showed an interest in buying horses, it is perhaps not surprising that the contributions Richard made to the family book of recipes were meant for the stables rather than the stillroom. When Anne and Richard's daughter Elizabeth came of age, she also began to add to the family collection. For example, a recipe for nosebleed and another for "looseness" were both signed "probatum Elizabeth Glyd."[24] "Probatum," probably short for the commonly used efficacy phrase *probatum est*, can be seen throughout the Glyd/Brockman collection; for example, the veterinary recipes Richard provided are also marked "probatum."[25] Elsewhere, in connection with a recipe to make a digestive "surfet water," "proved by my selfe" has been crossed out and replaced by "probatum Anne Glyd."[26] The annotations of the Glyd/Brockman recipe book suggest that Elizabeth, Richard, and Anne, much like the actors outlined in earlier chapters, worked collectively in collating and "proving" medical and veterinary know-how.

Among the Glyd daughters, it was not only Elizabeth who participated in the family's gathering and testing of recipes. Martha also remained close to her family and, even after her marriage to Richard Drake, continued to offer new recipes to her mother and her sister Anne Brockman. Much like the Fairfax women featured in chapter 1, Martha probably came into contact with new acquaintances and sources for recipe knowledge through her marriage. The know-how received from these new nodes within the network of recipe exchange was carefully notated. An example is a recipe for scurvy recorded as "this is my daughter drakes receipt of my Lady Clayton."[27] Additionally, a number of recipes connected with the Drakes appear in the loose recipe collection, including one for scurvy dated 1693 and instructions to make "*Elixer Proprietatis*" from a Sarah Drake.[28] Like her older sister Martha, the youngest Glyd daughter, Anne, continued to work on the recipe collection after she married William Brockman. Anne and her mother seem to have remained close after her marriage, which accounts for the Glyd notebook's surviving among the Brockman papers. One loose recipe, addressed to Anne Brockman while she was staying with her mother at Bletchingley in 1699, implies that the two worked together to augment their collection.[29]

The Brockman women also looked for recipe knowledge outside their immediate family. Examples include the recipe "for the piles for any bruise" that is signed "From my sister Chandler probatum Anne Glyd," the recipe to ease a hard labor from "my Aunt Eavens," and the recipe to make "surfet water" from "My Cosen Smith."[30] Years later, even after her mother's death, Anne Glyd Brockman continued to exchange recipes with the Eavens family, since the loose recipe collection includes a recipe for mouth water from "Cosen Ann Eavens," dated 1698.[31] As the individual recipes offered by various family members are written into the recipe book, so is a record of their participation and their contributions to the collaborative making of recipe knowledge. In this way, the Glyd/Brockman recipe books took on multiple functions. Aside from storing the family's recipe know-how, they recorded their collaborative knowledge-making activities, documented marriage alliances made by various family members, and mapped the extension of family networks.

Ledgers of Social Credit and Debt

Tucked into the Brockmans' collection of loose recipes is a sheet of paper addressed to "Mrs Scott at Longage to be left att Mr Fenners Coffehouse in Canturbury Kent" and later labeled "For the Jaundice Mrs Holder 1711." Instructions for making a purge composed of rhubarb, saffron, succory water, and syrup of roses and a powder made using eggshells, saffron, nutmeg, and sugar, purported to be good for "faintness," are outlined on the other side of the page. Somewhat curiously, at the end of these two recipes the writer strayed into rather different matters. He or she continued:

> Mr Richard Morgans wife is dead with a cancer in her breast Mrs Hammonds husband is lately dead and has left her 500 a year and a great deall of money she is taken a new house and she is about buying a pair of new coach horses Mrs Holder says she hopes her cosen does not, doe all this in order for a second match seeing now she has her liberty.[32]

In a flash the reader is transported from the stillroom and sickbed to the nucleus of domestic sociability: the early eighteenth-century parlor where gossip about friends, family, and kin was eagerly sought and exchanged. This document, in part a letter and in part a recipe, highlights the various social mechanisms framing the exchange, transfer, and codifying of household knowledge and demonstrates that recipes were readily tossed about alongside updates on the "liberty" and the health of various acquaintances. In earlier chapters I demonstrated the key role sociability played in recipe exchange and argued that recipes took an active role in contemporary economies of gift and obligation. Being given a recipe required reciprocating either with other recipes or with other favors. To give another example, when Sir Hans Sloane wanted to cultivate a friendship with Cassandra Willughby, daughter of the naturalist Francis Willughby (d. 1672), he first wrote on behalf of a mutual patron, the Duke of Montagu. His second letter, though, offered two favors: news that Sloane had successfully proposed Cassandra's brother Thomas as a member of the Royal Society, and a recipe for cashew sugar that Cassandra had enjoyed at the Montagu house. Sloane's gift of the cashew sugar recipe, Lisa Smith has noted, must have carried rich layers of meaning to Cassandra. The central ingredient of the recipe—cashews, a rare and important nut, referred to Sloane and Willughby's shared interests. The casual note that this was the very recipe Cassandra had enjoyed at the Montagus' further reminded her of their connections through a powerful and wealthy patron.[33] Added to this, as we saw with Conway and Harley and Sir John Tracy's ale recipe in chapter 3, sharing a recipe was a convenient way to open up continuing conversations and cultivate new alliances and friendships. For early modern men and women, then, it was crucial to keep track of this economy of recipe exchanges, and the manuscript recipe book did just that.

One important feature of such recipe books, as I outlined in chapter 1, is that they include the names of the recipe donors and sometimes when and where the recipe was obtained. Householders plainly felt this information was central to the record, and many went to great lengths to give detailed provenance. For example, Philip Stanhope, Earl of Chesterfield, described a particular medicine for gout as originally from "an Italian Doctor" but noted that he himself had received it from "Mrs Jane perrot of Lumberstreete, an acquaintance of my Lady Hubberts and by her often approved."[34] Another instance can be seen in Elizabeth Freke's remembrances and recipe collection. Writing in 1707, Freke was careful to note that a recipe for a palsy water was from "Lady Freke which now goes by Lady St John," but that the recipe was originally from Lady Freke's Dorset-based grandmother.[35]

Stanhope's and Freke's careful noting of the original "author" of these recipes emphasizes that many of them had a convoluted past and arrived in the compiler's hands through a series of intermediaries. More important, though, recording the names of these intermediaries was common. Various reasons likely fueled this practice. First, as we saw in earlier chapters, recipe makers had a tendency to return to donors with questions, so it was crucial to know who had given the information. Second, as is illustrated repeatedly, recipe collections were used across generations. The detailed source citation thus gave future generations and users information on the trustworthiness of the original donor. Finally, recalling that recipes were often exchanged as little presents—tokens of regard, affection, and gratitude—writing down the donors also acknowledged and documented presents received, and thus favors owed. Of course, the size of the favor was probably not fixed until the compiler could determine the recipe's efficacy and usefulness.

If we consider singular recipes as these small tokens, we may also read recipe collections as ledgers of gifts received. Thus they served both to track credits and debts in a social and knowledge economy and also to map family networks and alliances.[36] Consequently, as new generations of readers and users inherited the family book, they received not only treasured practical knowledge but also a written record of established family links and a perception of their place within the local community. To each family member reading and using the recipes, the collection acted not only as a reminder of social debts but also as a resource—a register of experts to be called on in case of need. In that sense household recipe books have much affinity with other contemporary writing practices such as the *album amicorum*, or friendship album, made popular by learned humanists, or the "papers" of seventeenth-century naturalists, such as John Aubrey's *Naturall Historie of Wiltshire*, which Elizabeth Yale has argued was "a site at which Aubrey could memorialize his friendships within the community of learned naturalists."[37] As records of a family's "wealth" in social connections and alliances, recipe books make perfect bedfellows with financial and legal papers, and as we will see in the next section, they were often archived together.

Chests of Papers, Books of Information

Alongside official and legal documents such as land deeds, wills, and records of leases, rents, and fines, the Brockman family kept, in the large wooden chests, a series of genealogical and family records, thus creating an archive not only of what they owned but also of who they were. Taken together, these papers record the family's land and estate holdings, their public role within the county, and the quotidian experiences of the household. The Brockmans were not unusual in creating a family archive. From the seventeenth century onward, English gentry put increasing emphasis on creating private archives preserving a written record of the family's pedigree, legal, and financial records in chests or in purpose-built muniment rooms.[38] More modest or curated versions of such archives were also created in bound notebooks. Here, well aware that their children and grandchildren would peruse them for years to come, the creators carefully crafted their volumes to include pertinent information ranging from legal and financial records to lists of births and deaths to informal maps of family relationships to their own hopes and dreams. A particularly lavish example is Cassandra Willughby, Duchess of Chandos's, "An Account of the Willughby's of Wollaton," which spanned two manuscript volumes.[39] In many cases these notebooks were also the repositories for the family's recipe knowledge. For early modern householders, it seems, recipes and family archives went hand in hand.

The Napier family of Holywell, Oxfordshire, provides such a notebook. The leather-bound book has the initials of the original owner, Elizabeth Powell (d. 1584), stamped on the front cover. Likely the book was brought to Oxfordshire when Elizabeth married William Napier (d. 1621/22). After her death the book passed through the hands of two of her sons, Christopher and Edmund (1579–1654), into the possession of her grandson

George (1619–1671) before finally ending up in the house of her great-granddaughter Margaret Napier Nevill.[40] Each generation of the Powell/Napier/Nevill family left its imprint on the notebook. Edmund Napier asserted his claim with the inscription "Sum Liber Edmundi ex dono Christoferi Napperi." The same hand wrote down his and his brother William Napier's birth dates and the date of Elizabeth I's coronation; these are followed by a record of Edmund's death written in a different hand. On another page is a list of twelve Napier children, likely the offspring of Edmund and his wife, born from 1610 to 1631. Two of the twelve, William and Elizabeth, died in infancy. For the others, no dates of death were recorded, but the family was careful to note down both the times of their births and the names of their godparents.[41] A generation later, Edmund's son George Napier and his wife, Margaret Arden, recorded important life events in their family. We learn that they married on 13 January 1647 and she gave birth to a boy, who sadly did not survive, about 7:00 a.m. Thursday, 30 August 1648. Margaret, Mary, and Frances followed in 1649, 1650, and 1652.[42] The notebook also contains miscellaneous financial and estate information such as leases, rents, and accounts relating to the estate of the Napier family and copies of letters and a list of books dating to the 1720s and 1730s.[43] Interspersed among the financial and legal information and family histories are scores of recipes, including instructions on how to preserve oranges, damsons, and other fruits and remedies to cure ague or fits or to make ink without heat or sun.[44]

Considered as a unit, this notebook presents an accounting of the Napier family's financial, personal, social, and intellectual lives over generations. Though on a smaller scale, the notebook served the same function as the two paper-filled chests Phyllis Brockman found. Crucially for my story, both the Brockman and Napier families considered recipe knowledge firmly part of the family archive. Just as recipe books were preserved in the same chest as land deeds and marriage settlements, recipes were written alongside financial accounts and family histories.

The miscellaneous information of the Napier and Glyd/Brockman family papers recalls the earlier Italian genre of *libri di famiglia* or *ricordanza* or the French *livre de raison*—family notebooks in which the patriarch recorded dates and citywide events affecting the family and entered copies of financial and legal documents.[45] Functioning as family archives, these books were closely guarded and passed down the patrilineal line, designed to perpetuate a written account of the family's fortunes or, as Natalie Zemon Davis puts it, "of the family arrow in time, of the careers and qualities of the parents, the training and marriages of children and the near-escapes and losses."[46] By creating and passing on a written record of a family's history, parents and elders were able not only to construct a family identity but also to articulate a strategy to ensure that the family's wealth—material, social, and financial—was maintained. While the transmission of the earlier *libri di famiglia* or *livre de raison* might have been strictly patrilineal, by the sixteenth and seventeenth centuries in England, women also often were keepers of genealogical information.[47] As Katharine Hodgkin has argued, early modern English gentlemen and gentlewomen used writing family histories and genealogies as a means to "affirm their status through the construction and reconstruction of family trees, connected ideally to continuous possession of land, displaying connections to prominent families and asserting the merits and gentility of progenitors." In the early modern English context, women were just "as zealous in the pursuit of pedigree as men [and] increasingly became family genealogists, remembering and recording the familial and domestic past."[48] As records of family participation and contributions, ledgers of social credits and debts, and maps of households' social networks, recipe collections fit closely into the work of writing family histories. The gendering of who was responsible for this complex task may, like other household paperwork such as financial accounting, follow the idiosyncrasies of individual families and respond to changes in responsibilities and needs at different life stages.[49]

Recipes and the Paperwork of Kinship

In constructing and inscribing their family histories in recipe collections, early modern householders used a range of paperwork and paper tools, including copying, compiling indexes, making lists of births and deaths, and keeping detailed records. In chapters 3 and 4 we saw how householders used paper tools and information management strategies such as multiple notebooks, ruled pages, and systems of marginal annotations to sort, categorize, and organize recipe knowledge as part of a multistep process of knowledge production centered on testing and trying. In those cases I emphasized the role of paper and note taking as methods through which householders signaled the epistemic status and value of individual recipes. As I argued there, householders and learned men shared similar information management strategies and paper tools. Like Francis Bacon and Robert Hooke, for example, Peter Temple and Elizabeth Godfrey used a series of neat books and waste or rough books to designate the epistemic status of knowledge. I thus firmly located household knowledge making and the related paperwork within the historiography of learned and largely scientific note taking.[50] However, paperwork and paper tools also played crucial roles outside university-based learning and were employed across early modern Europe as ways of enacting governance, managing complex financial accounts, and, on a more personal level, crafting and constructing identities through life writing.[51] While these studies tend to focus on the effect note-taking strategies had on the practices of early modern men, literary scholars such as Margaret Ezell and Victoria Burke have argued that early modern women, too, enthusiastically wrote in manuscript notebooks and used these paper strategies to make sense of their position within their worlds.[52] In what follows, I explore paper strategies used by household collectives, working across gender boundaries, in making recipe knowledge. Surveying a variety of practices, I argue that owing to the many meanings afforded to recipes and recipe books, men and women used a range of strategies under "the paperwork of kinship" to carry out their family and archival plans. This section focuses on three of these practices: recipes as family property, writing recipe books for one's children, and recording births and deaths.

Recipes as Family Property

In fashioning a collective "family book," Maurice Johnson was not alone. Other early modern householders, mostly men, used a range of paper strategies, from bookplates to reorganizing recipes, to construct books of family recipe knowledge. Below I present two examples, the Godfrey/Faussetts and the Chesterfields, to illustrate that refashioning a recipe collection into a "family book" had important consequences.

The Godfrey/Faussett recipe books are a set of five notebooks and two files of loose papers now in the Wellcome Library.[53] The books were created by generations of the Heppington manor household and remained *in situ* until the 1990s. Pasted-on paper labels title Wellcome Library manuscripts 7997 and 7999 "Heppington Receipts volumes 1 and 3." The cover of manuscript 7998 also bears traces of a pasted-on paper label, now lost, titling it "Heppington Receipts volume 2." Taken together, these books are fashioned to represent the recipe knowledge of the community at the manor of Heppington. Yet these recipe books were not always titled so. Two of the three volumes bear traces of their original compilers and own-

ers, Catherine Godfrey (fl. 1699) and Mary Godfrey Faussett (1695–1761).
The cover of volume 2 bears very faint traces of the inscription "Mary
Faussett alias Godfrey 1721," and volume 3 has inscribed on the front flyleaf "Catherine Godfrey her book 1698/9" and "Mary Faussett 1741."[54]
All three books also bear the bookplate of the Rev. Bryan Faussett (1720–
1776). In his mother, Mary's, book, Bryan's bookplate almost completely
obscures her initial ownership note (fig. 5.2).

Bryan might treasure his mother's collected medical and culinary know-
how, but he was rather quick to paste over her contribution to the volume and firmly assert his ownership.[55] To further push the point that these
recipes belonged to the entire Godfrey/Faussett family, a family member
(presumably Bryan of bookplate fame) added an index to three of the six
volumes. Titled "Index to the First/Second or Third volume of the *Heppington* Receipts" (my emphasis), these three indexes serve not only as an
information retrieval device but also as a means to tie the recipes to the
family seat and to the collective efforts of the entire Godfrey/Faussett
household. The knowledge these books contain was subsumed into the
general property of the family. Since family seats and their surrounding land, such as Heppington, were usually passed down the male line,
this knowledge once collated by the Godfrey/Faussett women also became the property of the Godfrey/Faussett men. The individual women
who worked to compile the knowledge were thus relegated to a secondary role, and gendering of the recipe knowledge shifted. Here the very
simple and common paper tools, the bookplate and the indexes, reveal
gender-related tensions and power dynamics within early modern households and in the creation of historical archives.

A collection of recipes associated with the Stanhope family is a second
example where paperwork and knowledge organization fashioned the la-
bors of individual women into a collective "family book."[56] The collection
was an ambitious project associated with Philip Stanhope, first Earl of
Chesterfield (1584–1656). The collection is titled

> A Booke of severall reseyts for severall Infirmities both in Man and Woman,
> and most of them, either tryed by my selfe or my wife, or my mother, or ap-
> proved by such persons, as I dare give creditt unto that have knowne the Ex-
> perement of it, themselves.[57]

Although only two of the handsomely bound folio-sized volumes have
survived, it is evident that these two were merely the beginning of a much

FIGURE 5.2. Wellcome Library, London, MS Western 7998, inside cover.

larger project. The collection is arranged alphabetically, and the surviving volumes cover the letters A–C and D–G. Within each letter section, recipes are listed in no particular order. For example, on the first page of the D section we find in sequence these ailments: dropsy, dropsy, delivery, dropsy, and deafness. Optimistic that the family will continue to gather know-how, blank pages were left in each letter section for future additions. Each section also ends with space for an index, which was completed only for the section D–G. While a number of hands can be detected in the manuscripts, the two volumes appear to be fair copies, and most of the text is written in two clear, practiced hands. As they stand, the volumes seem to be part of a bigger project to organize and categorize the existing medical knowledge of the Stanhope family. One can only conjecture what the original arrangement and notebooks might have looked like, but in this new scheme the female members of the Stanhope family appear only in secondary roles, much like the Godfrey/Faussett women. Even if they once owned or created their own books like other women I have described, they are now listed only as part of a family collective of recipe testers. In these instances creating a "family book" was also a way to merge individual investigations under the family name.

Writing for Your Children

Keen to advise their offspring and shape their lives, early modern parents offered their expressions of hope, good wishes, and a touch of parental influence through a number of contemporary textual genres. Some mothers embraced the "mother's legacy" genre as a means of shaping their children's beliefs and behavior.[58] Fathers also directed their guidance to future readers by compiling commonplace books of aphorisms. The commonplace book here served both as a way to account for the compiler's own experiences and also as a resource for future generations.[59] Alongside these genres, recipe books were another space where mothers and fathers encoded advice alongside other forms of knowledge. One demonstration of this practice is the intriguing inscription on the title page of Folger manuscript v.a. 430:

> Mrs. Ann Granvilles Book
> which I hope shee will make
> a better use of then her mother
> Mary Granville[60]

Under this inscription is written "Now Anne Dewes/Bradley 8ᵗʰ Sept 1740." Mary Granville's note hints tantalizingly at her guilt over her own lack of interest in housekeeping, or perhaps her skill. Yet at the same time it betrays this mother's sense of her responsibility to guide her daughter in the assigned role of housewife and household manager.

Fathers also hoped their daughters would make good use of the household knowledge passed on to them. For example, Sir Peter Temple of Stantonbury in Buckinghamshire, whom we met in chapters 3 and 4, also compiled a general medical guide for his daughter.[61] Temple's notebook for his daughter Elinor included both a collection of medical recipes and instructions on how to approach and use the material he provided.[62] These instructions comprised information on the origin of the recipes and the trustworthiness of the authors, a brief general method for preparing medicines, and short explanations of the way the book was arranged. Temple's persistent or perhaps reassuring voice can be heard throughout in his short notes of advice and endorsements of particular remedies. Consequently, through creating a collection of recipes to give to his daughter, Temple continued to shape and influence Elinor's behavior—how she maintained her own and her family's health, and how she managed her household affairs. In these cases, to inherit the family recipe book was to gain not only a cherished and prized set of know-how but also the expectations, hopes, and implied demands of one's parents.

On 2 October 1610 another father, the widower Valentine Bourne, inscribed his name in what would later become a miscellany of medical texts and family and local history.[63] Over the next twenty years or so, Bourne added not only medical, veterinary, and culinary recipes but also other relevant medical information including conversion charts of weights and measures, a section on the different kinds of pain taken out of Giovanni da Vigo's (1450?–1525) *The Most Excellent Workes of Chirurgerye* (London, 1543), a glossary of difficult medical terms, a treatise on urinoscopy, an extract from Galen's *De Alimentorum Facultatibus*, and a section on preserving health.[64] As the Napier family did, besides the medical information Bourne recorded both family and local history. On 1 April 1636 he wrote out the dates of birth, death, and marriage in his family from his own birth in 1566 to the death of his father-in-law in 1626. This record of his family's history was specifically "written for his loving daughter Elizabeth Bourne."[65] He also copied out lists of the mayors and sheriffs of Norwich from 1403 to 1648 and of all the high sheriffs from the reign of

Elizabeth I to 1660.[66] Within the recipe portion of the book, he left blank spaces at the end of each section, perhaps hoping his daughter would carry on his interest in gathering medical information and expand the collection with new recipes. Or it may just be that the book was still an ongoing, unfinished project.

"A Memorial of Our Childrens Birthes"

A number of other compilers or family collectives of compilers also paired records of births and deaths with their treasury of recipe knowledge. Deep in the back of one of the Brockman family recipe books is a list titled "A memorial of our childrens birthes: Sep: 22: 1650."[67] Underneath, listed in detail, are the names and birthdays of Anne Glyd's eight children: John, Elizabeth, Anne, Richard, Martha, Lawrence, Elizabeth, and Anne. For each entry Glyd carefully noted the child's date and year of birth as well as the day of the week. We learn that from 1650 to 1658 Glyd was continually pregnant and that the Glyds welcomed a child almost annually. Sadly, two of these children, the first Elizabeth and Anne, did not survive beyond their first few weeks. Furthermore, Lawrence and Richard both died as children, and the second Elizabeth died at just twenty-four. Anne's husband, Richard, also died in 1658, when all the children were still young. With so many deaths in the family, perhaps it is not surprising that Glyd continued to record the births of her grandchildren and her youngest daughter Anne's marriage to William Brockman. In each entry recording the death of a child, Anne, keeping meticulous track of time, recorded the child's age to the week. On the passing of her adult children, John and Elizabeth, Anne was careful to acknowledge God's grace in aiding both to be good, gracious, and "useful . . . to man." On the birth of each of her grandchildren, Anne noted the time of birth, expressed her gratitude to God, and put forward her hopes for the child's future. Consequently, on receiving her mother's notebook, Anne Glyd Brockman inherited not only her mother's medical knowledge and a copy of her family records but also expressions of her mother's hopes and dreams for her children and grandchildren.

Anne Glyd was not alone in combining records of births and deaths with recipes, although the Brockman papers reveal that the Glyds were avid writers of this kind of information.[68] A number of other families adopted this practice, marking and recording important life events.

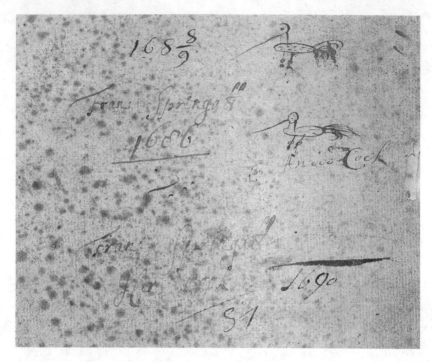

FIGURE 5.3. Wellcome Library, London, MS Western 4683, fol. 1r.

Another gentlewoman, Frances Springatt Ayshford, also took the time to write down information on her children's births. The lists of births and of deaths bookend her recipe collection, which also served as a record of Frances's life stages. Like many early modern gentlewomen, Frances began her recipe book while still unmarried and scrawled "Fran Springatt 1686" and "Fran Springatt Her Book" on the title page of the notebook. Later, Frances or someone else returned and added two ink drawings of a woodcock to the page (fig. 5.3).

On the verso of the ownership page, under a recipe for sticking "cheny" together with black snail slime, is a list titled "Daniel and Frances Ayshfords Childrens Born," followed by nine entries.[69] Like Anne Glyd, Frances Ayshford was pregnant or a new mother almost continuously for the greater part of a decade. The Ayshfords' first child, Francis, was born 12 August 1691 and was quickly followed by Adam, Edmund, Springatt, Thomas, John, Mary, and Daniel, who was born 1 July 1701. Richard, the youngest, was born 25 July 1709. Somewhat somberly, "dead" is

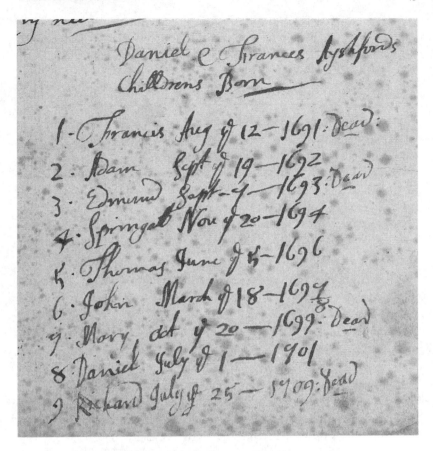

FIGURE 5.4. Wellcome Library, London, MS Western 4683, fol. IV.

written in a different ink next to Francis, Edmund, Mary, and Richard
(fig. 5.4).

Corresponding to the list of births is a list of deaths. On the verso of
the back flyleaf, listed under a recommendation for Mr. George Wilson's
"Anti Rhematick Tincture," is a list titled "Children dead of Dan and Fran
Ayshford." Here, listed chronologically, are the deaths of the four Ayshford children. Mary and Richard died before their third birthdays, Francis
was a little older, and Edmund died "at sea" at age twenty.[70]

Elizabeth Bertie Widdrington also included a list of her children alongside her recipes. Written in a beautiful decorated script and taking up two
folio-sized pages, the list is titled

An account when the Children of
the Right Hon:^{ble} William Lord Widdrington
by the Lady Elizabeth his Wife was borne
being all Baptized
the eighth day.[71]

The list includes fourteen children born from 12 November 1654 to 5 May 1675: Bridget, William, Mary, Elizabeth, Anne, Anne, Peregrine, Dorothy, Ellenor, Mr Henry, Ms Sophia, Mr Edward, Mr Roger, and Ms Katherine. The last five entries were written in a different hand, and the reference to "Mr Henry" and so on indicates they probably were made by a servant rather than by Elizabeth herself. Record keeping in early modern households, we learn, was a collective and collaborative enterprise.

Given the number of examples surviving, the combination of family history and recipe knowledge is unlikely to have been mere coincidence.[72] Early modern men and women saw creating a family's recipe book and recording births and deaths in the family as related endeavors. The use of terms such as "memorial" or "account" to describe these lists indicates that these acts of writing served a dual function of remembrance and reckoning. Of course, this kind of information was not paired only with recipes. Extant manuscript miscellanies connected with female compilers suggest it was fairly common to write family records alongside other information. In England, by the sixteenth and seventeenth centuries, both men and women often acted as custodians of genealogical information.[73] Family records like these can also be found mingled with household accounts, rent receipts, copies of sermons, and in the margins of texts such as the family Bible, the Book of Common Prayer, books devoted to childbirth, and almanacs.[74]

Conclusion

Inspired by Phyllis Brockman's extraordinary finds at Beachborough in the 1930s, I have argued that in the early modern period codifying recipe knowledge and writing family records went hand in hand. Like the papers in the chests she and her helpers pried open, manuscript recipe collections included a range of family records, including notes of the family's social alliances, financial and estate holdings, and medical and culinary interests.

As such, collecting and writing recipe knowledge is closely situated within these other forms of household record keeping. Family members also saw their recipe books as a space to acknowledge, record, and further social connections and family alliances. This wider framing of recipe collecting and writing positions the making of recipe knowledge within larger family and archival strategies.

The various schemes for creating recipe books described in this chapter underline the vital role these books played in preserving and recording family history. Householders used simple paper tools such as copying, making indexes or tables of contents, and inscribing ownership notes to mark books of recipes as the property of an entire family. Parents saw the family book both as a way to guide their offspring and express their hopes for them and as a space to memorialize significant events in their families' lives. Conventionally, historians have emphasized the bequest of recipe collections down matrilineal lines and the exchange of recipes among female family members and acquaintances. For example, Jennifer Stine and Sara Pennell have pointed out that many of these collections were given as wedding presents or were brought by women from their parents' home to their new household.[75] While that was undoubtedly true in many cases, it was by no means the only way collections changed hands. As the manuscripts described above demonstrate, men were actively involved in creating medical knowledge for the next generation and in receiving knowledge from their parents.

The younger generation, whether male or female, placed great value on the family book. Many chose to carry on their parents' work and continued to see these notebooks as a space for recipes and family records. As we saw in earlier chapters, the dynamic and unstable nature of recipe knowledge opened up spaces for conversation and negotiation. The Godfrey/Faussett and Stanhope family cases demonstrate that recipe collections witnessed tensions over the proprietorship of knowledge.

In his work on early modern autobiographies, Adam Smyth argues that men and women employed a network of life-writing texts to construct their self-identities. Reminding us of the old adage "what one is depends on what one owns," Smyth shows that writing and recording financial accounts was an important space where life writers fashioned selfhood.[76] Within this framework, the act of gathering family papers and recipes could also be seen as one way a family constructed its collective identity. Writing recipes was a way of writing the idea of the family.

Thus far we have largely considered making recipe knowledge within the household and manuscript culture. Yet a rich and colorful culture of printed recipes also flourished in early modern England. The next chapter situates our story within contemporary medical print and examines how manuscript and print worked dynamically together to produce household recipe knowledge.

CHAPTER SIX

Recipes for Sale

Intersections between Manuscript and Print Cultures

In 1656 a new book of recipes was introduced to the English reading public. Titled *Choice and Profitable Secrets both Physical and Chirurgical*, the work offered readers hundreds of medical recipes that had been "preserved as so many jewels of great value by the most virtuous Dutchesse of Lenox," who was known for the "rare and excellent cures" performed during her lifetime. John Stafford, the book's producer, included a portrait of the duchess to further connect the work with Frances Stuart, Duchess of Richmond and Lennox.[1] The recipes were framed by a wide range of health-related information including a first-aid guide titled "Helps for Suddain Accidents," information on critical days, and notes on the medicinal virtues of common herbs. Stafford acknowledged that for printed recipe collections "these times afford many choice and excellent pieces."[2] Indeed, *Choice and Profitable Secrets* joined a crowded bookshelf of vernacular health literature filled with a dazzling array of printed medical recipe collections all purporting to offer select and at times secret advice from a range of practitioners.[3]

Alongside *Choice and Profitable Secrets* on booksellers' shelves were several titles offering the "secrets" of aristocratic women, including *A Choice Manual of Rare and Select Secrets in Physick and Chyrurgery* (1653), associated with Elizabeth Grey (1582–1651), the Countess of Kent, and *The Queens Closet Opened* (1655), offering the recipes of Queen Henrietta Maria. Both were instant successes and remained staples on booksellers' shelves and readers' desks until well into the eighteenth century.[4] The market was also filled with books bringing the know-how of physicians, surgeons, and "expert operators" to the English reading public. These included the physician Gideon Harvey's *The Family-Physician, and the House-Apothecary* (1676); the Hertfordshire surgeon Thomas Brugis's *The Marrow of Physicke, or A Learned Discourse of the Severall Parts of Mans Body* (1640), and advertisements like Francis Dickenson's *A Precious Treasury of Twenty Rare Secrets* (1649).[5]

Up to this point in the book the focus of the featured case studies—the recipe books of the St. Johns, the Brockmans, Archdale Palmer, the Bennetts, and more—falls strictly on what Maurice Johnson so aptly calls "family books."[6] I have emphasized the need to view manuscript recipe books as personalized documents—intimately tied to a particular household's own history, social and local networks, and personal experiences. Yet the many recipe titles issued by printers/publishers suggest that print also played a crucial role in transmitting and codifying recipe knowledge. Reaching beyond the confines of the early modern household, this chapter explores the intersections, commonalities, and differences between the manuscript and printed recipe collections of the period. I argue that manuscript and printed recipe collections were part of a web of knowledge-making practices sharing similar patterns of transfer and use.

The chapter begins with an overview of recipe collections printed in England in the long seventeenth century. I outline the main characteristics of the genre and contend that many printed recipe books shared the open and malleable qualities that defined their manuscript counterparts. Turning to source citations, one of the central practices of recipe collecting, I highlight significant differences between the printed and manuscript recipe books, shedding light on how the communication media shape knowledge transmission. In the second section of the chapter I investigate intersections between printed and manuscript recipe collections by analyzing reading practices, emphasizing the significant crossovers between the two. Not only can we trace popular recipes across manuscript and print, early modern readers and users engaged with recipes in both media in much the same way. The chapter concludes by outlining the contributions this case study makes to our current understanding of manuscript and print cultures.

Printing Household Recipe Knowledge

Medical books have featured strongly on the lists of book producers and sellers from the beginning of printing in England. Contemporary booksellers' shelves were filled with books offering health-related information, such as medical textbooks, pharmacopoeias, herbals, surgical handbooks, expositions of new medical theories, regimen guides, and much more. City dwellers were frequently greeted by advertisements pasted on posts and walls offering all sorts of wondrous drugs claiming to cure ailments from worms in children to cataracts. As many studies have shown, medicine became increasingly commercialized over the early modern period, and the growth of vernacular medical print was closely intertwined with these broader changes in the provision of medical care.[7] Medical print played a crucial role within the medical marketplace, and the purchase of a printed book of medical know-how needs to be viewed alongside the buying-in of other medical services.[8]

Within the boom in vernacular medical books, recipe collections telling how to produce medicines occupied a central place on printers' lists and booksellers' shelves.[9] During the seventeenth century, over two hundred medical recipe books were printed, with about sixty of these being "new" titles rather than reprints or reissues.[10] In material form these works varied a great deal, from tiny duodecimo editions to nicely presented quarto books to short pamphlets of a couple of dozen recipes. On average the books probably cost from sixpence to one shilling sixpence depending on size and length, quality of binding, and paper used.[11] This would make the cost of a printed recipe collection similar to the cost of play texts but about three times the price of shorter, cheaper, and more broadly circulated texts such as almanacs.[12] As I will discuss below, there was also a lively secondhand book market that would have made many of these books available to generations of readers.[13]

In the first half of the seventeenth century, particularly in the 1610s and 1620s, the market in printed recipe books was dominated by a handful of favorites. These included John Partridge's *The Treasurie of Commodius Secrets* (1573) and *The Widowes Treasure* (1582), and Hugh Plat's *Delightes for Ladies* (1600) and its companion volume, *The Closet for Ladies and Gentlewomen* (1602). These works, together with another popular title from this period, Thomas Lupton's *A Thousand Notable Things, of Sundry Sortes* (1579), formed the bulk of the medical recipe book publishing in this period. Many of these titles remained in print for decades, with Lupton's *A Thousand Notable Things* reprinted in various guises and expanded editions until the nineteenth century.[14] In the 1630s and 1640s a few new titles joined these reprints, including Philbert Guybert's *The Charitable Physitian with the Charitable Apothecary* (1639), B. M.'s *The*

Ladies Cabinet Opened (1639), and *An Alphabetical Book of Physicall Secrets* (1639).[15]

Like many other genres of vernacular print, with the loosening of print censorship in the aftermath of the English Civil War, printed medical recipe books witnessed a boom in production.[16] More than fifty titles were printed in the 1650s, in contrast to the dozen titles printed in the 1640s and about twenty in the 1630s and 1660s.[17] Reissues of tried and tested titles continued to dominate the market, but *which* titles were reprinted changed rather dramatically. As other scholars have pointed out, the 1650s saw a new generation of recipe titles, many fronted by aristocratic female authors.[18] Elizabethan titles such as Partridge's *Treasurie of Commodius Conceits and Hidden Secrets* and *Widowes Treasure* and the extremely popular *A Closet for Ladies and Gentlewomen* all saw their last reprints in the 1650s. New issues such as *A Choice Manual* and *The Queens Closet Opened* rushed forward to replace them on the best-seller list and stayed there for the rest of the seventeenth century, a trend further explored in this chapter. In the 1660s and 1670s, while *A Choice Manual* and *The Queens Closet Opened* continued to dominate the market, they were joined by other popular titles, many of which continued the 1650s trend of packaging recipes as the "secrets" of noblemen and noblewomen. Examples include the "gentleman and traveller" William Lovell's *The Dukes Desk Newly Broken Up* (1660) and *Approved Receipts, or the Queens Representation* (1663).[19] George Hartman, a former steward of Sir Kenelm Digby, also became a prominent author of the genre. Hartman offered five recipe collections including three volumes of Digby's recipes and two collections under his own name: *The True Preserver and Restorer of Health* (1682) and *The Family Physitian* (1696). Another important author of the genre in this period was Hannah Woolley (or Wolley).[20] The wife of a schoolmaster, Woolley began a prolific career in the 1660s with the publication of *The Ladies Dictionary* (1661) and *The Cook's Guide: or Rare Receipts for Cookery* (1664), but her most successful book was *The Queen-like Closet* (1670), which was printed four times, with a second edition, *Supplement to the Queen-like Closet*, issued in 1674. Many of these titles became firm favorites with householders, who keenly read and annotated their own copies of the printed book and wrote adapted recipes into their own "family books."

From the early sixteenth-century offerings, many of the frequently reissued printed recipe collections were directed toward female readers and wealthy "ladies and gentlewomen," including Hugh Plat's *Delight for La-*

dies, *A Choice Manual*, *The Queens Closet Opened*, and *The Closet for Ladies*; John Partridge's two offerings, *Treasurie of Commodious Conceits* and *The Widowes Treasure*; and *The Accomplish'd Ladies Delight*. Other titles targeted those who could not afford other types of medical care. For example, Ralph Blower, the publisher of *A Rich Store-house or Treasury for the Diseased*, claimed that the work was "very necessary, and convenient to bee used of the poorer sorte of people (for the preservation of their health) that are not of abilitie to go to the phisitions."[21] This sentiment was echoed in works produced in the mid-seventeenth century, such as *The Poor-mans Physician and Chyrurgion* (1656), by Lancelot Coelson; Nicholas Culpeper's issue of Jean Prevost's *Medicaments for the Poor, or Physick for Common People* (1656); and Abraham Miles's *The Country-mans Friend* (1662). It is likely that there were motives besides the charitable behind these gestures. By addressing the collections to those who could not afford medical care, especially the attentions of a physician, the producers of these books ensured that their offerings would avoid the taint of commercialism and fit neatly and unthreateningly into existing structures of medical care. Concurrently, the producers of printed recipe collections likely also took part in more general schemes to open access to knowledge and, in the case of figures like Nicholas Culpeper, to render the learned physicians' knowledge accessible to all.[22] Mary Fissell has demonstrated that a relatively high percentage of authors of vernacular medical books in the 1640s to 1660s claimed to write "for the public good." In the early eighteenth century this trope was taken over by what Fissell terms "commercial books" (books that appeared to be promoting the author's wares).[23] This drive to publish for "public benefit" can be detected in a number of printed collections from the 1650s, such as Ralph Williams's *Physical Rarities* (1651), which repeatedly stressed that it was published to benefit the commonwealth.[24]

Reflecting the organization and working methods of the contemporary London book trade, most of the printed medical recipe books represent the collective efforts of a number of book producers, who often took the liberty of adding new material, reassigning authorship, or combining existing texts to create new titles. The book with which I opened this chapter, the Countess of Lennox's *Choice and Profitable Secrets*, is a good case in point. When the London publisher John Stafford launched the title in 1656, it was not the first public release for this set of recipes. In fact, *Choice and Profitable Secrets* was a revised, retitled, and slightly enlarged edition of a set of recipes that had graced booksellers' shelves for decades. These recipes first emerged in 1639 as *An Alphabetical Book of Physicall Secrets*, written by Owen Wood. The collection offered five hundred recipes, listed in alphabetical order, and "A brief Collection of all Hearbs, Plants, Seeds, Spices and Gums, now used in Physick."[25] Little is known about Wood, who was described as "no stranger in the practice of physic" and does not appear to be the driving force behind the publication. Rather, it was the book producer Walter Edmonds who, "having beene enformed of so much, was not willing to keepe the Treatise within private walls."[26] Perhaps to lend authority to the work, Edmonds invited Alexander Read, a contemporary medical practitioner who had published widely on surgery, to write a preface.[27] Although Edmonds himself never reissued the *Alphabetical Book* or, indeed, published any other medical text or recipe collection, these same recipes were issued twelve years later by another printer, T. B., as *An Epitomie of Most Experienced, Excellent and Profitable Secrets appertaining to Physick and Chirurgery* (1651). Although the title suggests it might be a digested "best of" version of the earlier publication, *An Epitomie* offered content almost identical to the *Alphabetical Book*. In this printing, though, perhaps in a bid to fool potential buyers into thinking the volume presented new material, the book producers shuffled the sequence of the first ten recipes. Like *An Alphabetical Book*, *An Epitomie* did not sell well enough to warrant a second edition.

Consequently, when John Stafford repackaged the same set of recipes as *Choice and Profitable Secrets*, he was merely following his predecessors in trying to gain a foothold in what was by the mid-1650s a fairly crowded market. Whereas T. B. only ventured to shuffle a few recipes around, Stafford took an altogether more radical approach. He assigned a new author, interspersed seventy new recipes in the original sequence, and appended Bradwell's tract on "Suddain Accidents."[28] The work evidently found some popularity with readers, since a second edition was issued in 1658. In 1660 Stafford offered English readers another new recipe collection, titled *A Manual of Most Experienced, Excellent and Profitable Secrets appertaining to Physick and Chirurgery*. In what is now a rather familiar story, it may be no surprise to find that the recipes had been offered to the public at least once or twice or three times before. No substantial changes had been made, and since the preface refers to it as the fifth edition (which the 1658 edition would have been), it is doubtful that this was even a new copy. In other words, over two decades a single set of recipes appeared under four titles in seven editions, issued by three publishers and connected with at least two authors.[29]

The print history of this set of recipes illustrates the fluidity and instability of vernacular medical print in mid-seventeenth-century England. As
Adrian Jones and others have argued, book production was highly collaborative, with authors, compositors, proofreaders, printers, and booksellers all shaping the final product.[30] Book producers both created and
responded to market demands, reprinting and repackaging books accordingly. It was not uncommon for them to take liberties in assigning and reassigning authorship, combining different titles and texts, adding to and
modifying a book's contents, and reissuing old copy.[31] We might draw par-
allels between household and print shop collectives. In both cases the collaborative knowledge making, involving multiple hands, brought with it
an open attitude toward assigning authorship.

John Stafford's introducing the "Countess of Lenox" as an alternative
author of the *Choice and Profitable Secrets* also brings to the foreground
one of the most significant changes to the printed medical recipe genre
in the 1650s: titles fronted by female compilers and authors.[32] As Patricia Crawford has written, there was a huge leap in books by women in
the second half of the seventeenth century, with a peak in the 1640s and
1650s. Female authors appeared in print in increasing numbers and over a
wide range of topics, from political pamphlets to prose fiction to medical
recipes and advertisements.[33] Within the printed recipe genre, as I noted
above, the 1650s offerings associated with both Elizabeth Grey, Countess of Kent, and Queen Henrietta Maria gained popularity with readers.[34]
Other collections fronted by aristocratic women include *Natura Exenterata* (1655), which was seen as the "Experiments" of Alethea Talbot Howard, Countess of Arundel (c. 1584–1654),[35] and *Queen Elizabeths Closset
of Physical Secrets* (1656), which purportedly presented

> two other physicall pieces, one of them collected by a great Navigatour, of his
> own Experiments, and presented with his own hands, to our late Queen Eliza-
> beth; the other being a Physitians Collections, drawn with his own hands from
> an antient Manuscript found in an Abby at their dissolution, with some of his
> own Observations and Experiments annexed thereto.[36]

While one can only guess which Elizabethan physician is referred to here,
one assumes that some of these recipes may have originated from the
collection of Sir Walter Raleigh, who was well reputed for his cache of
recipes. Hence, although a woman graced the cover of the book, it was
learned men who penned the contents and devised the experiments. This

tactic of drawing on the authority of both aristocratic women and learned
men can also be seen in Stafford's repackaging of *Choice and Profitable
Secrets*, where he was careful to retain the connections with medical men,
stating that the recipes had been "perused and methodized" and that "severall eminent and able Doctors have both approved and (for the good of
the Republick) recommended it to the Presse . . . so the whole Nation may
have the benefit of it."[37] Stafford was not shy in naming these "eminent"
medical men: Owen Wood, Alexander Read, a Dr. Johnson, and Steven
Bradwell, author of the appended tract on "suddain accidents." Thus, in
just a few sentences Stafford managed to position his offering as a parcel
of secrets connected to a noblewoman and also as know-how backed by
learned doctors.

   It is important to note that these collections were all posthumously attributed to aristocratic women rather than actively written and published
by countesses, queens, and duchesses. As to their sharing their ideas in
the press, it is difficult to separate the original intent of Elizabeth Grey or
Queen Elizabeth or the mysterious Countess of Lennox from the machi-
nations of the book producers. In fact, as Elizabeth Spiller has noted, even
in the cases of *A Choice Manual* and *Natura Exenterata*, where contemporaries often identified the female authors with the texts, the women
named on the covers of these volumes might not have in "any conven-
tional sense" written the texts they contained. The printed work might be
a version of their households' manuscript recipe books, or the manual of
recipes might have been composed as a gift for the named noblewoman,
much like the examples in chapter 1.[38]

   The sudden surge in printed recipe collections associated with aristocratic women issued during the Commonwealth period is not, of course,
devoid of political meaning. As happened with many other genres of vernacular print, the political turmoil in the 1640s and 1650s prompted a
boom in the production of medical books. Recently scholars have argued
that during the 1650s book producers used the humble recipe collection
as a polemical tool, and books such as *The Queens Closet Opened* served
as a continual reminder of the royal family during the Protectorate.[39]
These texts also encouraged readers to make the connection between
good household management and the return of the Stuart monarchy as
the "head" of the national household.[40] Concurrently, by marketing the
book as once belonging to a famous woman, the booksellers relied on the
social clout of the original (purported) compiler to carry the volume and
implied that the recipes offered were in use within her household. In pur-

chasing the book of Elizabeth Grey or Frances Stuart or Alethea Talbot Howard, readers were also catching a glimpse of the domestic practices within aristocratic households at large.

This focus on practice and the use of recipes within particular domestic settings emphasizes the important place of experiential knowledge. Much as with the manuscript recipe books discussed in earlier chapters, personal observation and experience were paramount in establishing authority. For example, in the case of *Queen Elizabeths Closset of Physical Secrets*, the "navigatour" presented the recipes to Elizabeth with "his own hands," and the physician also copied his contribution with "his own hands" and supplemented it with "his own Observations and Experiments." Salvator Winter, author of *A New Dispensatory of Fourty Physicall Receipts* (1649), boasted to readers that his recipes were collected out of his "manifold Travels through Europe, Asia and Africa," where he "observed and learned many secrets."[41] In *Natura Exenterata*, the pseudonymous author of the preface, "Philiatros," argued that "practical observations are more assistant then *Systemes*, obscure words rather beget amazement then amendment in sick people. They who *do* (though emperically) are to be preferred before those who dispute and talk." Contending that "*Garrulus medicus est onerosior morbo*" (a garrulous doctor is more burdensome than death), "Philiatros" claims that the recipes in the work were recommended "because they have done upon many."[42] The authority of these remedies, then, was based on their having been widely tried and tested rather than on book learning. This privileging of empirical knowledge was widespread in vernacular medical print across Europe by this period.[43] Yet here privileging empiricism also locates the publication within the ongoing medical controversies in early modern London.

As many scholars have noted, throughout the mid-seventeenth century university-trained physicians struggled to gain and maintain authority in a diverse and complex medical marketplace.[44] The College of Physicians launched several attempts to exert control over medical practice and publishing and was embroiled in disputes over the reform of both medical regulation and theory. The debates were partly rooted in the college's desire to safeguard its monopoly, but they also reflected intellectual changes in the medical field, where empirical experiment was gaining ground against theoretical argument based solely on the study of ancient medical texts.[45] Some of these conversations, both the call for reforms and the college's responses, were carried in print. Against this backdrop we might view calls by "Philiatros" and others to privilege experiential knowledge within these debates.

The multiple changes of author on the title page of *Choice and Profitable Secrets* indicate that book producers and sellers, at least, saw the title page as the main way consumers assessed the book's recipe knowledge. In this the case of *Alphabetical Book/Choice and Profitable Secrets* is not unusual. In her study of Hannah Woolley's cookbooks, Margaret Ezell has argued for the importance of the "name (and face) on the title-page."[46] In fact, the book producers placed so much emphasis on the author listed on the title page that they often omitted information on the authors of individual recipes. *An Alphabetical Book of Physical Secrets*, for example, recorded the name of the recipe donor in only two out of nearly 590 recipes, and in *A Choice Manual* only about a dozen individual recipes out of over three hundred were associated with a recipe donor.[47] This trend for omitting the authors or donors of individual recipes can be observed in most printed collections produced in the period. John Partridge's *The Treasurie of Commodius Conceits and Hidden Secrets* (1573) provided author information for only one of the 70 recipes offered, and *The Accomplish'd Ladies Delight* (1675) named only four authors for the 109 recipes in the text. Similarly, the producers of Nicholas Culpeper's *Culpeper's School of Physick* and Gervase Markham's *The English Hus-wife* did not feel the need to associate any additional names with the individual recipes in the book. In both these cases Culpeper's and Markham's popularity with readers undoubtedly pushed sales.

Of course there are exceptions to this general trend. Three titles associated with aristocratic compilers and courtly cultures all offered substantial information on the donors of individual recipes. These are *The Queens Closet Opened* (1655), *Natura Exenterata* (1658), and a translation of Oswald Gäbelkhover's *Artzneybuch* as *The Boock of Physicke wherein throughe commaundement of the most illustrious, and renoumned Duke and Lorde, Lorde Lodewijcke . . . Most of them selected and approved remedyes* (1599).[48] All three collections originated in aristocratic households and lived previous lives as manuscript "family books," and it may be that the book producers deemed including famous recipe donors a selling point. Certainly in the case of *The Queens Closet Opened* they emphasized original recipe donors by offering a composite list of names cited at the beginning of the book.[49] Yet even here the names of recipe donors were attached to only about one-fifth of the recipes. *The Boock of Physicke* is particularly interesting. The German editions of Gäbelkhover's work were wildly popular, and the printed work made a treasured manuscript recipe collection from the court of Duke Ludwig of Württemberg available to the reading public.[50]

The work, however, was not as popular in England as in Germany, suggesting that while German readers might be attracted by the recipes circulating at the Württemberg court, English readers showed rather less interest.

Aside from these three titles, it is clear that the producers of printed recipe collections did not place much weight on individual recipe donors. The emphasis in most printed recipe books was on the author or compiler of the entire collection. This practice, of course, contrasts sharply with the manuscript recipe books. Earlier chapters have stressed the importance of family, neighborly, and social networks in codifying recipe knowledge in manuscript. In this context, whether documenting family histories or accounting for social credits and debts, source citations played a central role. Recall from chapter 1 that the notebooks of Archdale Palmer and the Bennett family were filled with rich, detailed information on where, when, and from whom recipes were collected. For the creators and owners of manuscript notebooks, writing down the origin of a recipe was an integral part of collecting recipe knowledge.

Consequently it is clear that, although they circulated contemporaneously, the printed and manuscript medical recipe books differed in the crucial area of source citation.[51] The printed recipe books, framed by a confluence of market needs and political ambitions, tended to focus on presenting the compiler of the collection as an authority. Collectively and collaboratively compiled over a long time, manuscript recipe books, while recording individual ownership of the book, tended to focus more on the authors of single recipes. These differences are tied to circulation patterns and the differing social and economical contexts shaping the two forms of communication.

In general, manuscript compilers retained control over access to their collections by restricting copies and transcriptions.[52] Printed recipe collections could be bought by any reader, either new or secondhand. At least one author, Robert Boyle, was anxious about the wide dissemination of these texts and the lack of control over just *who* had access to them. Published in 1692, Boyle's *Medicinal Experiments, or a Collection of Choice and Safe Remedies, for the Most Part Simple and Easily Prepared* contains one hundred recipes organized into decades. The first five decades were printed in 1688 as *Some Receipts of Medicines, for the Most Part Parable and Simple, Sent to a Friend in America*.[53] In both volumes, as in many other printed collections, the names of the recipe donors are left out. Michael Hunter has argued that Boyle withheld the names of the contributors to protect their commercial medical practices.[54]

In their place Boyle offered his readers two systems of his own devising in the 1688 edition. The first separated the recipes into three classes or orders. "A" denoted recipes that were the most reliable and were recommended as "very considerable and efficatious in its kind," "B" represented the second, lesser sort, and "C" recipes were of the lowest order.[55] Boyle provided further testimony of his own experience with these recipes, writing, "And because I presume it may be expected, that I should give you some short Character, of those whereof I have had some kind of Experience; whether by the Relations of such as have been Cur'd, or much Reliev'd by them; or by my own Exhibiting or Prescribing them; or by Reports receiv'd from those that engag'd to try them or that employ'd them upon my account." Since he remained reluctant to cite authors or provide anecdotal evidence, Boyle offered an additional system of 1 for one trial, then crosses and stars, to denote how many times each recipe had been tried within his knowledge, though, as Michelle DiMeo has noted, this secondary system was not included in *Some Receipts* or in *Medicinal Experiments*.[56] Neither of these systems achieved wider use by other producers of printed collections; however, they echo practices householders used to annotate the trying and testing I described in chapters 3 and 4, suggesting shared expectations of recipe trials across knowledge communities.

The social and economic contexts shaping the production and use of manuscript and printed recipe collections also had an effect on author citation. As I discussed in earlier chapters, manuscript recipe notebooks had multiple roles within the household. They were at once books of family histories, accounts of gifts received and exchanged, and ledgers of social credit and debt. Information on from whom and to whom knowledge was given was thus crucial. Additionally, the sociability of recipe exchange gave name citations a purpose aside from indications of expertise or authority. As friends and neighbors browsed through one's collection and appreciated its contents, the names cited must have also highlighted one's social milieu and perhaps strengthened one's position within it. As family members, particularly younger generations and extended kin, consulted the "family book," the names cited would undoubtedly have triggered memories (fond or otherwise) of particular family members and reminded readers of family connections that both required continual cultivation and could be used to further the fortunes of present family members.

These deeply embedded social meanings afforded to "family books" meant that manuscript and printed recipe collections circulated and were distributed in distinct ways. As with many other textual genres in manu-script, "family books" often remained more or less within the same social circles.[57] Although compilers appear to have ventured outside for single recipes, collections tended to stay within the family, passed down from generation to generation, and at times within the same four walls. Con-sequently, inheriting a recipe collection must have entailed inheriting a social circle, so the names of the authors within the collection would be recognizable to future readers. Seeing the same names but lacking the in-herited social knowledge, however, the new owners of a printed collection would not be able to reconstruct the social circle of the original compiler. In these cases the names cited would have little meaning, since many of the reasons for citing names in manuscript collections were negated once the printed collection was taken out of its social context.

The printed recipe collections sold by London booksellers lacked the social meanings afforded to handwritten notebooks. Unlike manuscript "family books," these were not unique products tailored to the needs and interests of a particular family or household. Functioning somewhat like the "starter" collections described in chapter 1, printed recipe books pre-sented quick and ready practical information, available for sale to anyone whose pockets were deep enough. If, as we saw in earlier chapters, family and social credit shaped access to recipe knowledge, in the realm of print the currency for recipe exchange was hard cash. This shift in economics from obligation to monetary changed the mode of access to recipe knowl-edge and also the readership.

Reading and Using Printed Recipe Collections

Printed recipe collections, alongside other vernacular medical texts such as general herbals, pharmacopoeias, books of simples, general medical guides, and regimen books, provided householders with a constant source of information. Providing they could gain access, these books consider-ably extended the scope of recipe compilers. This section reconstructs the way householders read printed collections to fuel their recipe knowledge.

Past readers have long enjoyed the attention of historians and literary scholars alike, and studies on the history of reading have argued that the act of reading is highly individual and culturally conditioned; each read-ing thus makes meaning.[58] Reading can be studied in several ways, includ-ing re-creating contemporary libraries, examining secondhand book sales, investigating marginal notations, and looking at manuscript notebooks such as commonplace books. People's reading notes illuminate their encounters with books and illustrate occasions when they did particular types of reading and the strategies they used to record these experiences. Many recent narratives of reading and note-taking practices have mainly concerned the learned sphere. Lisa Jardine and Anthony Grafton have introduced us to Gabriel Harvey; Kevin Sharpe wrote about William Drake; and William Sherman reconstructed the notes of John Dee.[59] These case studies demonstrate that many early modern learned readers "read for action," and some even were "professional readers," reading and digesting texts for their patrons.[60] Less attention, however, has been paid to reading by less educated readers.[61] Heidi Brayman Hackel and Lori Humphrey Newcomb have admirably filled the gap by exploring women's reading and leisure reading, and they describe readers' interactions with works of literature such as Sir Philip Sidney's *Arcadia* and Robert Green's *Pandosto*.[62] Studying how householders engaged with vernacular medical print and recipe collections extends our current understanding of early modern reading beyond learned and literary writings. Concurrently, the focus on reading everyday print and manuscript recipe books complements recent studies on reading and household science.[63] Framing recipe knowledge within the practicalities of everyday household management encourages us to view these activities as "reading for practice."

As readers, householders engaged with printed recipe collections in familiar ways.[64] They claimed ownership by inscribing their names in books; wrote and drew systems of signs such as crosses and check marks in the margins; and copied lengthy extracts and summaries into their manuscript notebooks. When these are studied alongside many of the manuscript-based knowledge-making and information management strategies described in previous chapters, it is clear that strong continuities existed across manuscript and print cultures. To illustrate and showcase these continuities, the next section focuses on three points of intersection: annotated margins, traces of print in manuscript and manuscript in print, and crossover recipes.

Annotating Printed Recipe Collections

Studying annotations and marginalia has proved to be a rich path into the everyday practices of early modern readers. As the pioneering studies by Grafton, Jardine, and Sherman have demonstrated, careful consideration and deciphering of marginal notes can open a window on our historical actors' interactions with specific genres and their reading goals at particular moments.[65] Studying marginal scribbles in printed recipe collections proves especially fruitful both in illuminating householders' attitudes toward these books and in teasing out the connections and close relations between print and manuscript cultures. Many surviving copies of printed recipe books contain traces of readers' engagement with the text.[66] These traces can be separated into three kinds of marks: ownership notes, selection marks, and corrections or amendments.

Just as they might sign their names with a flourish in their manuscript recipe notebooks, many owners of printed recipe collections inscribed their names in the books. These owners include William Milne, who scribbled his name several times throughout his copy of the 1666 edition of Ralph Williams's *Physical Rarities*, and Narcissus Luttrell, who bought and inscribed a copy of the 1675 edition of *The Accomplish'd Ladies Delight*.[67] The ownership notes indicate two trends about readers and printed recipe books. First, many of the ownership notes show that these collections remained treasured long after their initial publication dates: William Milne inscribed his copy of *Physical Rarities* more than three-quarters of a century after the work was published.[68] Other examples include Mary Mott, who on 29 June 1703 wrote her name in a copy of the 1655 publication *Natura Exenterata*, and Dorothy Hawkins, who in 1668 wrote hers in a 1630 edition of *A Closet of Ladies and Gentlewomen*.[69] Second, as Luttrell's note indicates, although large portions of the book producers' offerings targeted gentlewomen, they were not the only ones who found printed recipe collections interesting. The *Queens Closet Opened* is another book that, though addressed to female readers, attracted both sexes. Copies of the book in the Folger Shakespeare Library include a 1698 edition with the signatures of John Frances, dated 1704, and Samuel L[oi?] mis, dated 1726.[70]

Additionally, like their manuscript counterparts, printed recipe collections frequently changed hands, and many have multiple ownership notes. A copy of *A Hundred and Fourtene Experiments and Cures of the Famous Phisitian Philippus Aureolus Theophrastus Paracelsus* (1583) is now in the Tanner Collection in the Bodleian Library. Before Thomas Tanner (1674–1735) acquired it, the book had two owners: Robert Cook, who inscribed his name on 10 August 1587, and Robert Estrake.[71] Printed recipe books were also shared within the family. For example, a copy of the 1562 edition of *The Secretes of the Reverend Maister Alexis of Piemont* has several ownership notes, including "Mary Holmes her booke anno dom 1640"; "George Holmes his book anno"; and "John Holmes his booke anno dom 1640."[72] Finally, another work that frequently changed hands is Sir Kenelm Digby's *Choice and Experimented Receipts in Physick*. The Bodleian Library copy was once in the collection of John Locke but had been transferred to William Elliot by 18 January 1778.[73] Another copy in the Folger Shakespeare Library was shared by members of the Heywood family, including Thomas Heywood, who signed his name in 1695, and Francis Heywood, who inscribed his in 1672, 1692, and 1699.[74] Since Thomas's and Francis's notes overlap in time, one wonders whether the book's ownership was once contested. Recipe knowledge had a long shelf life, whether in manuscript or in print.

Annotations and marginal notes in such volumes also demonstrate that many printed recipe books were heavily used. For example, the reader and annotator of a copy of Thomas Brugis's *The Marrow of Physick* (1640) recorded his efforts at producing medicines by writing "made this" next to certain recipes.[75] Thus we know that he tried both the recipe for syrup of dried roses and the one for syrup of damask roses.[76] Another reader wrote "distill this" or "distilled this" by many recipe titles in a copy of Conrad Gesner's *The Newe Jewell of Health* and apparently found the "golden water" "an excellent water of lyfe for myselfe to use."[77]

Other kinds of marks show that, perhaps not surprisingly, users of these books and collectors of recipe knowledge approached the printed texts in much the same way as their manuscript counterparts. That is, the recipe knowledge London book producers offered often went through the same multistep trials and testing described in chapters 3 and 4. In print as in manuscript, recipe users assessed and tested recipes in myriad ways. Some compilers used common efficacy phrases to warn or inform future readers of their own experience with particular recipes. Much like the Okeover family we encountered in chapter 3, a reader of *A Closet for Ladies and Gentlewomen* wrote the letters *b* and *g*, presumably for "bad" and "good," next to certain recipes.[78] A reader of the 1579 edition of Thomas Lupton's *A Thousand Notable Things* liked to write "this is true" or "you shall finde it true" at the end of particular recipes.[79] Another reader wrote "this hath beene experimented by many persons of quality" next to a recipe for a drink against the plague and the pox in a copy of Thomas Brugis's *The Marrow of Physick* (London, 1640).[80] Finally, another annotator used the efficacy tags "proved" and *probatum* throughout a copy of *The Secrets of the Reverend Maister Alexis of Piemont*.[81]

As in the manuscript texts, readers also altered dosages, substituted ingredients, and changed production methods to reflect their own preferences. For example, the reader of *The Marrow of Physick* mentioned above was apparently happy to substitute "muscadine" for "malmsey" and add "mithridate" in a "recipe against the plague, pox, measells and other infectious diseases." Additionally, he or she changed the dosage Brugis stated, increasing it from two spoonfuls to three.[82] Readers also put in extra instructions reflecting their own experience with working through the recipe. The annotator of a copy of *The Queens Closet Opened* (1658) added an extra step to a recipe for syrup of gillyflowers, straining the liquor after letting the gillyflowers stand in water for twenty-four hours. In another recipe "to dress a pig," the annotator reminds future readers to slit open the pig before skinning it as described in the recipe.[83] Along the same lines, some annotators explained particular materia medica. For example, in a copy of *A Choice Manual* a reader annotated the famous recipe for the Countess of Kent's powder by underlining "lapis contra parvam," placing an *X* in the margin.[84] On the facing page is a two-page description and history of the ingredient "lapis contra yerve." The notes tell us it was first introduced into England by Sir Francis Drake, discovered on a Spanish ship where it was labeled as meant for the king of Spain. The drug was commonly used as an antidote for snakebite. Finally, on occasion readers expressed disbelief at certain recipes. In one instance the reader of *The Secrets of the Reverend Maister Alexis of Piemont* was vocal about his thoughts on certain recipes, and next to one "to make a light in the night" wrote that "non but fools will experience it."[85] Notations like these in printed books suggest that their owners intended to return to these recipes, either for rereading or for repeated use. If we position these books within working and reading household communities, we might also assume that their pages bear the traces of conversations between household members. It may be that the annotators were motivated to inscribe their thoughts in the margin not only as reminders for themselves but for future readers.

Readers of printed recipe collections also augmented their books. Many wrote in newly encountered recipes on the flyleaves of the texts, and some, though rarely, inserted recipes into the text.[86] Readers also extended the printed book by binding in blank pages. In these cases we could perhaps even think of these printed collections as being like the manuscript starter collections described earlier. A copy of Thomas Elyot's *The Castell of Helth* (1572) has been augmented with several pages filled with household recipes.[87] In another case, one particularly enthusiastic compiler bound an extra gathering of blank pages into a copy of Thomas Brugis's *Vade Mecum*, filling them with reading notes taken from *Secrets of the Reverend Maister Alexis of Piemont* and works by Nicholas Culpeper.[88] Doing so created a personalized medical compendium suited to individual interests and needs.

Readers thus treated particular copies of printed recipe collections much as they would a manuscript collection. They laid claim to both the material object and the knowledge it contained by inscribing their names. Multiple ownership notes gave individual copies of printed works a history and, much like the manuscript "family books," recorded the many hands through which recipe knowledge flowed. In the case of recipes, both print and manuscript were seen to be inherently malleable and open to additions and augmentations. Readers and users of printed collections selected, tested, and modified the recipes they contained and augmented the printed text with their gathered knowledge. By adding their own recipes, whether gained from reading other manuscripts or printed texts or orally from their family and social acquaintances, these readers personalized the printed texts and adapted them to their own needs.

Writing Print into Manuscript and Manuscript into Print

Aside from annotating printed recipe books in their libraries, a number of householders also incorporated selections, some brief and some extensive, from printed recipe collections into their own manuscript recipe books. Much as learned humanists approached their reading notes in diverse ways, recipe readers' note taking also took various forms, including digests or summaries, notes organized under "heads," or reading notes acting as external indexes.[89]

Since most creators of surviving household recipe books were elite men and women, many had the money to buy printed medical books and often used them in conjunction with their family books of recipes. An excellent case is the paper archive of Margaret Boscawen, a gentlewoman living in late seventeenth-century Cornwall, whom we met in chapter 1. Boscawen and later her daughter Bridget Fortescue created several manuscript recipe books of various shapes and sizes and a substantial collection of recipes written on loose slips of paper. Alongside this complex paper archive, it appears that they also read and consulted a range of printed medical books, including Elizabeth Grey's *A Choice Manual, The*

Queens Closet Opened, Nicholas Culpeper's translation of the *Pharmaco-poeia Londinensis*, and Alexander Read's *Most Excellent and Approved Medicines* (1651), as a way to enlarge the family's recipe repertoire.[90] Traces of Boscawen's engagement with these books can be seen in many of her notebooks, including her large folio recipe book and a long, slim "plant" book with information on gardening and plant storage. In many cases Boscawen's reading notes record only the book title, recipe title, page number, and her own experience with the cure. This is particularly evident in her list of "excellent waters," with entries such as "A soveraign water good for many cures: to be stilled in a limbacke; [title in shorthand] 44" or "A water to comfort weake eyes and to preserve the sight: 49: Lady Kent. It may be stilled at any time in an ordenery Rose still:" or "to make synamon water 3 or 4 severall wayes in the London Disspensitory: 140."[91] These notes might be sparse, but they demonstrate the role printed books played in Boscawen's recipe practices. Her recommending specific equipment such as rose stills or limbecks hints at experience with producing these waters. And Boscawen's inserting this list within her manuscript recipe collection suggests she viewed the printed know-how as part of her usual medical practice. Finally, these sparse notes imply that all four books were part of Boscawen's library and that her reading notes let her find the recipes quickly and easily.

Though Boscawen used her reading notes as an external index to her collection of printed medical books, other compilers, who might not have had continuing access to printed books, took rather longer notes. Elizabeth Freke, for example, made selective but substantial notes from John Gerard's *The Great Herball*. Freke's notes focused on the medicinal "vertues" or uses of particular herbs, thus forming an ideal complement to her medical recipe collection.[92] In another example we see an anonymous eighteenth-century reader create a personal "digest" of *A Choice Manual*. This slim notebook presents dozens of recipes taken from the work, selected and reorganized according to the reader's needs.[93] In both these cases readers were not merely copying information from printed books but rather were reading recipes, perhaps assessing them on paper, and selecting those they deemed worthy of trial and suitable for their families. In this their engagement with print might be seen as analogous to the recipe "collecting" stage described in chapter 1. Whereas some householders like Archdale Palmer gathered recipes around dinner tables and at taverns, others, perhaps more bookish or more remotely situated, turned to print for new ideas.

Some readers of printed recipe collections categorized and sorted their reading notes, giving them easy access to the information in the future and sketching out their ideas of the body and health. Two manuscripts are clear examples of this. One is anonymous but perhaps associated with a "cousin Greenway," and the other is associated with Thomas Sheppey. Both Greenway and Sheppey took careful notes from a number of printed medical books such as Lazare Rivière's *The Practice of Physick* (1655) or Nicholas Culpeper's translation of the *Pharmacopoeia Londinensis*, which was first published in 1649 as *A Physical Directory* but was revised in 1653 as *The London Dispensatory*.[94] In both, the collected information was pieced together to produce one unique volume under a newly devised organizational method based on subject categories reminiscent of a contemporary note-taking practice: commonplacing.[95] Here, though, the headings are the names of ailments and body parts, offering future readers easy access to the knowledge. Through collecting and reorganizing the recipes under new sets of headings, existing texts were taken apart and transformed into new ones, which were constantly changing according to the interests and needs of the compilers.

While the numerous examples offered thus far document the transfer of recipe knowledge from print to manuscript, it is clear that knowledge also traveled the other way. If we take printers and publishers at their word, many of the well-known collections began life as manuscripts. Gervase Markham, for example, claimed the recipes offered in the *English Hus-wife* were taken from "an approved Manuscript which he happily light on belonging sometime to an honorable Personage of this kingdome, who was singular amongst those of her ranke for many of the qualities here set forth."[96] Likewise, W. M., who put forth *The Queens Closet Opened* in 1655, stated in his preface that most of the recipes presented in the printed work were at one time transcribed into Queen Henrietta Maria's book by his own hands, and that he had kept many of the "original papers." It was only when he encountered two manuscript copies of the said recipe collection abroad that he decided to dispatch his "original" copy to the press in a bid to prevent circulation of the false or corrupt copies.[97] In both these cases, of course, the book producers had much to gain by revealing the original manuscript status of their printed works, and the disclosure might well have been calculated to improve sales. However, we should not discount the idea that many printed collections likely existed in manuscript form before being modified and adapted to the new medium of print.

This brief look at the textual transfer from print to manuscript and back shows that it was a road well traveled. Householders, I have argued, used the varied offerings from the London book producers to augment and extend their collections of recipe knowledge. They consulted not only the numerous printed recipe collections available, but also cognate genres such as herbals, which highlighted the medicinal virtues of herbs, and observations or cases, which detailed specific instances of drug use. This dynamic relationship between print and manuscript demonstrates that on paper and in books, householders participated actively in commercially produced medical knowledge.

Crossover Recipes

Early modern householders' enthusiasm for consulting vernacular printed recipe collections is also reflected in the many "crossover" recipes evident across the recipe archive. Two recipes, the Countess of Kent's Powder and Paracelsus's Plaister, were particularly popular with recipe compilers and, owing to their distinctive titles, are easy to trace across the print and manuscript sources. In print the Countess of Kent's Powder made its first appearance in *A Choice Manual of Rare and Select Secrets* (1653). Of the know-how offered in the book, the Countess of Kent's Powder was particularly well-known and popular with readers and recipe collectors alike. The powder, supposedly good against "all malignant and pestilent diseases" including the French pox, smallpox, measles, the plague, fevers, and melancholy, was made from precious ingredients such as pearls, crabs' eyes, amber, hart's horn, coral, crabs' claws, oriental bezoar, saffron, ambergris, and lapis contra yerva.[98] The powder was well-known among diverse contemporary communities of knowers. In 1654, for example, Samuel Hartlib noted that the powder was highly recommended by Sir Kenelm Digby, who thought it had performed many "miracles." Digby reported that when the late King Charles I was sick with the smallpox, "the whole Coll of the Rojal Physitians advised the taking of this powder, which hee had no sooner taken but the sensible effects of his recovery appeared to the admiration of them all."[99] While the list of ingredients suggests that the recipe might be too expensive for everyday use, and indeed for most householders to make at home, it also often appears in the "family books."

Anne Glyd was one of many contemporaries whose collections contain a version of this recipe.[100] Glyd marked it as "proved Ann Glyd," as she did with many other recipes in the book, suggesting firsthand knowledge with the cure. Other recipe compilers who copied out versions of the Countess of Kent's Powder include Johanna St. John, Mary Dacre, and Susanna Packe.[101] Yet using recipes from printed books did not always take the form of copying out a recipe. Peter Temple recommended the Countess of Kent's Powder for an ague, noting "this I have often proved with good successe P. T.," without noting down the instructions for making it.[102] It may be that he had the instructions at hand elsewhere, or perhaps he bought the medicine ready-made from an apothecary.

Another set of instructions that gained similar popularity is a recipe known as Paracelsus's Plaister. The exact origin of this recipe, purportedly translated from Latin, remains to be discovered; one particular version, printed in *The Queens Closet Opened* (1655), appears in several manuscript recipe books. In print the recipe is titled "Paracelsus his Plaister cal'd Emplastrum fodicationum Paracelsi, good for many diseases herein mentioned, translated out of Latine into English." The multistep recipe required four kinds of gums (galbanum, Opoponax, Ammoniacum, and Bdellium), wine vinegar, olive oil, wax, and a host of expensive ingredients such as myrrh and frankincense.[103] If makers could obtain all the ingredients and worked through all the steps, they were to be rewarded with a medicine that healed sores, ripened impostumes and biles, and was excellent against "the Canker and Fistula, the Shingles or Saint Anthonies fire; and also a sovereign and speedy help against all pains, to asswage all aches and for all kinds of wounds . . . is a singular and special help for bones out of joynt."[104] Given all these uses for common maladies, it is not surprising that we find this recipe in several manuscript recipe collections, many already familiar to us, including the book of Anne Brumwich and the Fairfax family, the Johnson family book, the Okeover family book, and Johanna St. John's collection.[105]

In adding recipes gleaned from printed collections to their family books, some householders made fairly faithful copies, but in other cases they abridged, simplified, or otherwise modified the instructions. An excellent example is the copy of Gascon's Powder in the recipe book associated with Susanna Packe in the Folger Shakespeare Library. Gascon's powder is another recipe often copied from *A Choice Manual*. Like the Countess of Kent's Powder, the drug requires luxury ingredients such as pearl, coral, crabs' eyes, amber, and hart's horn and purportedly cured fever. Packe's version of the recipe is simplified in several ways. Most significantly, in *A Choice Manual* Packe merges two versions, only one using oriental bezoar, into a single recipe. She also pares down the production method and notes that here saffron is added for color. Finally, whereas the printed version advises the patient to take ten or twelve grains in "Dragon-water, Carduus-water or some other cordiall water," Packe instructs that one should give children six, eight, or ten grains according to their strength and that thirty grains is more appropriate for adults. Though Packe does not advise patients to take the powder in cordial water, she does add that they should take the medicine at the beginning of fevers or any malignant distemper.[106]

As explored in previous chapters, such frequent modifying and customizing was central to making recipe knowledge in the early modern household. Some of the manuscript versions of the two recipes tracked in this section are labeled as a particular person's "way." A recipe collection associated with the Okeover family has an entry "Paracelsus his Plaister," attributed to Lady Darcie, and the two versions of "Lady Kent's Powder" in Johanna St. John's book are attributed to Dr. Willis and to a "Mrs Archer."[107] Once modified and adapted, these personalized versions of popular recipes begin to circulate among manuscript books. The St. John family's two recipes for Lady Kent's Powder are a case in point, since they appear both in Johanna St. John's "great receipt book" and in a copy likely made by Johanna's daughter Anne Cholmondeley.[108] The concurrent circulation of the two versions of the recipe attests both to the value placed on personalized versions of well-known remedies and to the dynamic and continual transfer of recipe knowledge from print to manuscript.

Conclusion

> A Collection of above three hundred and fifty most excellent choice receipts in Physick and Chyrurgery and Preserving Conserving and Cookery and Pastry, *None of which have been ever published in Print but in Parcells Transmitted in Writing* to diverse Families. In Order to instruct their Posterity. By those worthy Persons in the next page.[109] [my emphasis]

So began an anonymous late seventeenth-century recipe book now in the National Library of Medicine. The book in question, probably a presentation copy, is a folio-sized notebook containing hundreds of recipes written in a clear, uniform hand. As hinted by the title, the following page offered a list titled "The Names of Those Noble Persons By whose Assistance I made this Collection." It names, in order of social rank, five titled gentlemen (of whom three were doctors of medicine), nine additional doctors, ten ladies, ten gentlemen, and sixteen gentlewomen.[110] This is followed by a list of commonly used weights, information on "opening roots," "emolient herbs," and the greater and lesser hot and cold seeds.[111] The recipes are separated into three parts containing first medical recipes, then recipes for sugaring and preserving, and finally recipes for cookery and pastry. In form this beautifully scribed manuscript clearly mirrors *The Queens Closet Opened*, which also began with a list of contributors (arranged by rank) and presented its contents in three parts, including "The Pearl of Practice," covering medical recipes; "A Queen's Delight," offering confectionary recipes; and "The Compleat Cook," covering culinary recipes.[112] This emulation of print by manuscript would be unremarkable were it not for the writer's claim that none of the recipes "have been ever published in Print but in Parcells Transmitted in Writing." Thus the recipes this manuscript contains are valuable because they had never seen the light of print, yet the form and method of presentation intrinsically connects them to vernacular medical print. This clever positioning by the writer encapsulates the complex relation between manuscript, print, and recipe knowledge.

Early modern English readers were offered a wide array of printed recipe books. Some promised a glimpse into the worlds of aristocratic women and even queens; others presented know-how gained on long, meandering travels across Europe; still others purported to be convenient digests of dozens of learned medical tomes.[113] Printed and manuscript recipe collections sat side by side in kitchens, stillrooms, or libraries as crucial parts of the household recipe archive. For many householders, manuscript and print worked together in the making and transfer of recipe knowledge.

By examining reading practices and the intersection between the two media, I have argued that manuscript and print cultures were in constant, dynamic dialogue. Printed recipe collections shared the open-ended, malleable qualities of manuscript texts, in contrast to other textual genres of natural inquiry such as natural history.[114] This openness can be viewed from various perspectives. First, book producers amended, extended, and merged recipe texts in much the same way as manuscript creators did. However, whereas manuscript creators viewed the accumulation of recipe knowledge within systems of social networking and family strategies, the producers of printed recipe collections likely had more commercial interests in mind, merging successful texts to create new versions of popular titles. Second, and perhaps more significant, readers and users saw both manuscript and printed recipe collections as adaptable and open to modification. The annotations and interlineal markings in both kinds of collections demonstrate that their users selected and tested recipes, modifying them to suit their needs and tastes. Recipes, whether in manuscript or in print, were sets of instructions to be used, tried, rewritten, and reused.

Despite these commonalities, if we return to the opening quotation we are reminded that at least one anonymous writer attached value and cachet to recipe knowledge that was found only in manuscript circles and that had "never been seen in Print." Part of this value, as many have argued, derives from the more controlled and limited circulation of manuscript texts, which led to a select audience.[115] This is to some extent due to the social and cultural meanings accruing to manuscript recipe books as material objects and as records of social alliances and family history. The omission of source citations in printed recipe collections reminds us that, for all its merits, on its own vernacular print did not embody these deep meanings. That is not to say that householders could not inscribe those meanings in a purchased printed book: as their ownership notes and additional bound-in pages filled with recipes attest, they frequently did so. Rather, printed recipe collections were part of other contemporary conversations. As I illustrated above, they promoted normative ideas about gender roles and gendered labor. Recipe collections written by women surged in the 1650s because this was exactly the kind of work expected of early modern women.[116] Concurrently, printed recipe collections, as a number of studies have shown, were also vehicles for pushing particular political viewpoints. Some might have been bemoaned their wider circulation, but it certainly gave others a space to disseminate much more than recipes.

The case of recipes thus prompts us to further consider early modern manuscript and print cultures. This book's strong focus on manuscript culture and knowledge making aligns it with recent studies by Lauren Kassell, Deborah Harkness, and Elizabeth Yale.[117] Like Simon Forman's casebooks, Clement Draper's and Hugh Plat's notebooks, and the papers of seventeenth-century naturalists, early modern householders' manuscript recipe books were sites of active knowledge making. By bringing these to the fore, I seek to position manuscript and scribal culture alongside the well-studied print culture of early modern science. Nonetheless, as this chapter has shown, manuscript and print were interconnected systems for producing and disseminating knowledge, each with its own set of values and meanings. Future studies of early modern knowledge might do well to further investigate these entangled connections.

CONCLUSION

Recipes Beyond the Household

Hardyman,

I received all things [ac]cording to your letter. Your master saith he will be at Lidiard this week but I beleve it wil be tuesday first but be not wholly unprovided for him and if he wil have a venison pasty get Mrs duell or my la newcombs cook to make for it. I would not have you troble mrs perkins about it. . . . I would have you git Mr goram to distil me this water the receit of which I have sent here and such of the things as are not to be had in the contry and fr the hearbs and the rest you must git such as goody wolford knows—not at an apothecar-yes and if ther be no hearts horn in the paper bags in the skullery buy it at marlburough . . . the glas stil [is] for Mrs giles perkins for I brok hers. Thank her for my chees and I thank your mother kindly also and have sent her Mr Jacombs funeral sermon.

Your loving Mrs Johanna St John

I would have the waters that are stild keept ther not sent nor this which is to be stild on which must be writ treacle water and taken in several pints the first 2 and third so as they are received if the cardus and redrose water can be got redy it would be most convenient to have it don . . .[1]

It seems apt to end this book with one final letter from Johanna to Hardyman. Between her fretting about Sir Walter's venison pasty, her mindfulness to replace Mrs. Perkins's glass still, and her enclosing a recipe to make treacle water, this letter encapsulates the wide range of concerns that regularly occupied Johanna. Recipes and recipe knowledge were framed by the complex concerns of everyday life. Early modern house-hold maintenance, it is clear, included looking after the "health" of the household in a number of ways—bodily, social, political, and financial. Putting food on the table and keeping her household healthy was merely one of many issues on Johanna's mind. For her, household management involved ensuring that her estate and workers produced precious distilled waters, venison, and French and Portuguese melons to offer as "little presents" and that the decorative gardens at Lydiard continued to reflect the refined tastes of the St. Johns. This meant making sure one always had on hand the right materials, equipment, expert helpers or technicians, and—importantly—knowledge. Recalling Johanna's panicked letters concerning Muscovy ducks for the king and pork and cheese for Edward Hyde's visit, it seems we might do well to place her efforts at household management alongside her political and social ambitions. Recipes played a central yet multifarious role in these schemes of household maintenance and management: they were at once tidbits of know-how, tokens of affection and regard, and ties to family histories and social networks.

The household as a site for making recipe knowledge, this book has shown, had far-reaching consequences. The main cultural frameworks operating within the early modern household—family and social—shaped recipe knowledge in a multitude of ways. For many fathers and mothers, writing it was combined with a dose of family history. The family's legacy of kinship and its hard-earned recipes were heirlooms to be bequeathed alongside hopes and dreams for the next generation. Recipe books were also records of social networks and, most crucially, accounts or ledgers of obligations and gratitude. Sharing a recipe, we have learned, required reciprocation and was the first act in an alliance, old or new. Writing down recipe knowledge, I have argued, was potently influenced by these frameworks of social and family strategy. Writing was at the very heart of domestic economies.

Yet the everyday, the domestic, and the social also had profound epistemic effects. The crucial function that recipe exchange served in forming and maintaining interpersonal relations means that recipes were requested, received, and gathered from a medley of characters and for a range of reasons. This inherent openness in these transactions meant that many householders approached recipe collecting as "bring it home first, decide what to do with it second." Within this model of knowledge making, assessing, testing, and trying recipes is central. It is thus perhaps predictable that a figure like Edward Conway used every pathway—from books in his library to personal observations to asking experts in his circle—to assess the gathered know-how. It is also not surprising that Peter Temple scrawled "quare" throughout his recipe notebooks. Codifying recipe knowledge was a complex affair.

Emerging from this study is a sense that the contexts and framing of household recipe knowledge created unusually open and malleable texts.

Often created and used over generations, household recipe books were scribbled over, marked up, crossed out, written and rewritten. This enabled methodologies borrowed from the lively fields of histories of reading and note taking to further elucidate the subject. By combing through marginal notes, interlineal interjections, Xs, and other scribbles, I was able to reconstruct and track methods of knowledge making. While some of these methods, such as "recipe salvage," might be exclusive to household recipes, I hope this book also offers a blueprint for codifying knowledge that can be adapted to other contexts. The multistep collating of knowledge and the myriad ways men and women tested recipes run parallel to practices in other epistemic spaces such as artisanal workshops or even in the more practically oriented projects of learned societies, such as the London Royal Society's History of Trades program or the early investigations into mineral spa waters by the medically oriented members of the French Academy of Sciences.[2]

In exploring recipe trials, I have argued that early modern households were vibrant scenes of knowledge making. Driven by curiosity and the real need for trustworthy know-how, householders were continually putting recipe knowledge to the test. In doing so they were also exploring the natural and material world, for our historical actors were not only testing for efficacy but carefully monitoring how materials reacted in particular environments, noting how individual bodies responded to remedies, and figuring out the most expedient or cheapest way to produce particular drugs or to apply a simple eye water. The rich details arising from these cases inspire us to reflect further on quotidian knowledge practices. Recipe trials, I have argued, were one of the main pathways through which householders gained deeper understanding of sickness and health, of the human body, of natural and man-made processes, and of materials.

Moving through the chapters of this book, we were also continually reminded that while gentlewomen and gentlemen ended up writing down household knowledge, they were aided by a team of men and women who also possessed hands-on experience and expertise with a range of practical tasks. In Johanna's opening letter we see that Mr. Goram was assigned the distilling, Goody Wolford was consulted on the ingredients, and the glass still was to be returned to Mrs. Giles Perkins. Specialized distillation equipment, then, was borrowed and returned, just as we might now turn to our neighbors for a certain size of cake pan or flan dish. Recalling Bess the dairymaid and Rudler the gardener, it is also clear that other staff members at Lydiard knew about botany and the life cycles of livestock.

Through domestic work, they engaged daily with natural knowledge. We might call this household science.

By recovering their story, this book seeks to add these rich, at times messy, practices into narratives of knowledge making. Doing so widens the cast of historical characters participating and contributing to early modern science. They are no longer "invisible technicians" but rather makers, writers, and transmitters of natural knowledge. Their practices and ideas enrich our current narratives of the Scientific Revolution not only by elucidating the very spaces and contexts in which men like Robert Boyle and Robert Hooke worked but also by extending the parameters of early modern natural inquiry. By all accounts, these men and women were filled with wonder and curiosity about the world around them, which they investigated and explored in a wide range of interconnected ways.[3] This book has shown that, rooted in practices of making and doing and reading and writing, household recipe practices entailed a complex web of different kinds of knowledge—about the natural and material worlds, production techniques and processes, social and family strategies, and health and the human body. Exploring recipes thus grants us a glimpse into the lives of early modern men and women and into the making of "everyday knowledge."

Acknowledgments

In a book about collective and collaborative knowledge making and the recording of knowledge encounters, it is an enormous pleasure to express my appreciation and gratitude to multiple scholarly communities, institutions, and friends. Over the years, conversations formal and informal with friends and colleagues have helped me make sense of an enormous and unruly archive and craft the story presented here. At Oxford, Margaret Pelling took the seedling ideas I brought with me and helped me develop them into a research project. Her never-ending support and critical feedback have shaped my ideas about medicine in early modern England in profound ways. I still have fond memories of her office with those boxes of index cards and maintain that the apprenticeship was all too short. Years ago, at Wellesley College, Katy Park introduced me to the history of medicine and science, encouraging me to dig into the stories of early modern women. I owe much to her teaching and her exemplary scholarship.

The long gestation period for this book and the whims of modern academic life put me in the enviable position of being part of several institutions and departments, all congenial environments for research and writing. I owe particular thanks to colleagues at the Wellcome Unit for the History of Medicine at Oxford, the history departments at Leicester and Warwick, and the Department of the History and Philosophy of Science at Cambridge. I am grateful to the Wellcome Trust, the Leverhulme Trust, Lincoln College Oxford, Wellesley College, the Folger Shakespeare Library, and the Huntington Library for their financial support over the years.

In 2012 I joined the Max Planck Institute for the History of Science (MPIWG), and it is there that most of this book was written. I am grate-

ful to the Max Planck Society for funding my research group, "Reading
and Writing Nature in Early Modern Europe." Like many scholars in the
field, I remain indebted to Lorraine Daston, who invited me into her department, welcomed me, challenged me with her astute questions, and offered sage advice at just the right time. Over the years, colleagues and
visiting scholars at the MPIWG have, through passing or prolonged conversations, inspired me to expand the scope of this project and to think
deeply about the place of recipe knowledge in histories of science. My
first thanks go to my dear colleague Christine von Oertzen. Her invitation to the "Beyond the Academy" working group led me to Berlin, to
conversations about gender, science, and technology, and to my joining
a group of scholars with unmatched intellectual generosity. Three other
MPIWG working groups have framed and shaped this project at crucial
junctions: "Structures of Practical Knowledge," "Testing Drugs and Trying Cures," and "Working with Paper." Those who have the opportunity
to read the resulting publications by these groups alongside this book will
easily recognize my indebtedness to those intense and constructive discussions. My thanks go to all members of these working groups and to
MPIWG colleagues past and present, but especially to Carla Bittel, Dan
Bouk, Montserrat Cabré, Sven Dupré, Sietske Fransen, Ursula Klein,
Sally Kolhstadt, Nina Lerman, Anna Maerker, Elena Serrano, Pamela
Smith, Claudia Stein, Viktoria Tkyaczk, Matteo Valleriani, Simon Werrett,
and Elizabeth Yale.

The interdisciplinary nature of this study brought me into contact
with a number of scholarly communities. Many chapters and ideas of this
book have benefited from feedback and discussions in seminars, workshops, and conferences on both sides of the Atlantic over the past decade.
I am grateful to the organizers for their invitations and hospitality and
to audiences for thoughtfully engaging with my work in seminar rooms
and at cafés and pubs. Lynn Botelho, Lorraine Daston, Michelle DiMeo,
Mary Fissell, Amanda Herbert, Cathy McClive, Christine von Oertzen, Alisha Rankin, Lisa Smith, Simon Werrett, and Elizabeth Yale have all read,
commented on, and improved individual chapters in the book. Lauren Kassell read multiple drafts of this book in its entirety. I appreciate her time
and am thankful for her interventions and support. Lauren's critical feedback, encouragement, and general pushing and prodding have been crucial
to my completing this book in its present state and at this present moment.

In 2012 Lisa Smith invited me to launch an academic blog on recipe
studies. Five years and hundreds of posts later, *The Recipes Project* has

become an interdisciplinary hub for recipe research. My thanks go to the entire *Recipes Project* community for their passion and enthusiasm for discussing all things recipe related and, by sharing their research, encouraging me to see connections across temporal and geographical boundaries and across fields of knowledge. I am also thankful to members of the Early Modern Recipes Online Collective who share my dream of making manuscript recipe texts more accessible and inspire me with their strength as feminist scholars.

My thanks also go to Karen Merikangas Darling and her team at the University of Chicago Press. Karen took me on as a first-time author and quietly and patiently held my hand through publication. Two anonymous readers carefully read and reviewed drafts of this book. Their perceptive questions and comments helped me to crystallize my ideas and fully develop my arguments. Later in the production process, Alice Bennett edited my manuscript with meticulous care. The readability of this book owes much to them.

Acknowledgments in a project about books and manuscripts (particularly reading practices) cannot be complete without thanks to our colleagues in libraries and archives. Staff members at the Wellcome Library, Bodleian Library, British Library, Folger Shakespeare Library, Huntington Library, New York Public Library, New York Academy of Medicine Library, and Glasgow University Library brought up manuscript after manuscript, printed book after printed book, always with a smile. Largely owing to them, the library is and perhaps always will be my "happy place." For the past six years I have had the privilege of working with the librarians at the MPIWG, so my thanks go to Urs Schoepflin, Esther Chen, and their team, particularly Ellen Garske, Ruth Kessentini, and Urte Brauckmann. Likewise, I am incredibly grateful to my student assistants—Julia Jaegle, Daniel Glombitza, and Mareike Hennies—for all their help over the past few years.

For their friendship and good cheer, and for listening to me go on and on about recipes, I owe an enormous debt to all my friends in and out of academia. You know who you are! Finally, my thanks go to members of the Leong and Ferries families. First, my immense gratitude goes to my parents and to my sister, Vicky, for their unwavering support of my dream to be a historian. I also thank Lilian, Peter, and Luk Suk for keeping me (and my scholarship) honest over the years. This book is dedicated to Alex and Nicholas Ferries, who have been there for me every step of the way. Alex has read every draft of this book, talked me through every research and writing hiccup, and, perhaps most important, distracted me from the book when I most needed it. Thank you and thank you again. Nicholas still giggles whenever cheesecake recipes come up in conversation, and even though he thinks this book can't possibly be nearly as interesting as Percy Jackson's adventures, he still wants to read it. Needless to say, all the flaws are my own and anything of interest owes much to them.

Notes

Introduction

1. Huntington, HM MS 41536, fol. 163v.
2. Ibid., fol. 165v.
3. The Derings' adventures in testing and trying are discussed in detail in chapter 3.
4. For an illustrative handlist and bibliography of seventeenth-century English manuscript and printed recipe books, see Elaine Leong, "Medical Recipe Collections in Seventeenth-Century England: Knowledge, Text and Gender" (PhD diss., University of Oxford, 2006), appendixes A and B.
5. For a detailed discussion of both these cases, see Mary Floyd-Wilson, *Occult Knowledge, Science, and Gender on the Shakespearean Stage* (Cambridge: Cambridge University Press, 2013), chaps. 1 and 4.
6. For more detailed discussion of household economies and management, see chapter 2. For an overview, see, for example, Karen Harvey, *The Little Republic: Masculinity and Domestic Authority in Eighteenth-Century Britain* (Oxford: Oxford University Press, 2012); and Amanda Herbert, *Female Alliances: Gender, Identity and Friendship in Early Modern Britain* (New Haven, CT: Yale University Press, 2014), chaps. 2 and 3. For the German context, see Alisha Rankin, "The Housewife's Apothecary in Early Modern Austria: Wolfgang Helmhard von Hohberg's *Georgica curiosa* (1682)," *Medicina e Storia* 8, no. 15 (2008): 59–76.
7. Mark Greengrass, Michael Leslie, and Michael Hannon, *The Hartlib Papers* (Sheffield, UK: HRI Online Publications, 2013), http://www.hrionline.ac.uk/hartlib, 13/206B-207A.
8. Mary Fissell, "Women, Health and Healing in Early Modern Europe," *Bulletin for the History of Medicine* 82 (2008): 1–17; Margaret Pelling, "Thoroughly Resented? Older Women and the Medical Role in Early Modern England," in *Women, Science and Medicine, 1500–1700: Mothers and Sisters of the Royal Society*, ed. Lynette Hunter and Sarah Hutton (Thrupp, Stroud, Gloucestershire, UK: Sutton, 1997),

63–88. Recent studies that showcase the early modern home as a site of medical activity include Hannah Newton, *The Sick Child in Early Modern England, 1580– 1720* (Oxford: Oxford University Press, 2012); Alun Withey, *Physick and the Family: Health, Medicine and Care in Wales, 1600–1750* (Manchester: Manchester University Press, 2011); and Olivia Weisser, *Ill Composed: Sickness, Gender and Belief in Early Modern England* (New Haven, CT: Yale University Press, 2015).

9. For an overview, see Andrew Wear, *Knowledge and Practice in English Medicine, 1550–1680* (Cambridge: Cambridge University Press, 2000), esp. chap. 2. Lucinda Beier's essay on the Josslin family provides a good illustrative case study: "In Sickness and in Health: A Seventeenth Century Family's Experience," in *Patients and Practitioners: Lay Perceptions of Medicine in Pre-Industrial Society*, ed. Roy Porter (Cambridge: Cambridge University Press, 1986), 101–28. See also Elaine Leong, "Making Medicines in the Early Modern Household," *Bulletin of the History of Medicine* 82, no. 1 (2008): 145–68; and Seth LeJacq, "The Bounds of Domestic Healing: Medical Recipes, Storytelling and Surgery in Early Modern England," *Social History of Medicine* 26, no. 3 (2013): 451–68.

10. My use of "everyday knowledge" here references Michel de Certeau, *The Practice of Everyday Life*, trans. Steven Rendall (Berkeley: University of California Press, 1984). I also reference recent conversations in the history of knowledge. For conceptual statements, see, for example, Peter Burke, *What Is the History of Knowledge?* (Cambridge: Polity, 2015); and Lorraine Daston, "The History of Science and the History of Knowledge," *KNOW: A Journal on the Formation of Knowledge* 1, no. 1 (2017): 131–54. For a historiographical overview of this emerging field, including its connections with Wissengeschichte, see Johan Östling et al., "The History of Knowledge and the Circulation of Knowledge: An Introduction," in *Circulation of Knowledge: Explorations in the History of Knowledge*, ed. Johan Östling et al. (Lund, Sweden: Nordic Academic Press, 2018), 9–33.

11. See, for example, essays in Elaine Leong and Alisha Rankin, eds., "Testing Drugs and Trying Cures," *Bulletin of the History of Medicine*, special issue, 91 (2017).

12. Hannah Woolley, *The Cook's Guide, or Rare Receipts for Cookery* (London, 1664); and Bartolomeo Scappi, *Opera dell'arte del cucinare* (Venice, 1570). See also Deborah L. Krohn, *Food and Knowledge in Renaissance Italy: Bartolomeo Scappi's Paper Kitchens* (Farnham, UK: Ashgate, 2015); and Gilly Lehmann, *The British Housewife: Cookery Books, Cooking and Society in Eighteenth-Century Britain* (Totnes, Devon, UK: Prospect Books, 2003).

13. Deborah E. Harkness, *The Jewel House: Elizabethan London and the Scientific Revolution* (New Haven, CT: Yale University Press, 2007), chaps. 5 and 6.

14. George Starkey, *Alchemical Laboratory Notebooks and Correspondence*, ed. William R. Newman and Lawrence M. Principe (Chicago: University of Chicago Press, 2004); and William R. Newman and Lawrence M. Principe, "The Chymical Laboratory Notebooks of George Starkey," in *Reworking the Bench: Re-*

search Notebooks in the History of Science, ed. Frederic L. Holmes, Jürgen Renn, and Hans-Jörg Rheinberger (Dordrecht: Kluwer, 2003), 25–41.

15. The literature on the recipes genre in general is fast growing and rich and constitutes too large a body of works to fully list here. The pioneering studies on such texts are John K. Ferguson, *Bibliographical Notes on Histories of Inventions and Books of Secrets*, 2 vols. (1898; repr., Staten Island: Pober, 1998); and William Eamon, *Science and the Secrets of Nature: Books of Secrets in Medieval and Early Modern Culture* (Princeton, NJ: Princeton University Press, 1994). For an overview, see Elaine Leong and Alisha Rankin, eds., *Secrets and Knowledge in Medicine and Science, 1500–1800* (Farnham, UK: Ashgate, 2011); and Michelle DiMeo and Sara Pennell, *Reading and Writing Recipe Books, 1550–1800* (Manchester: Manchester University Press, 2013). Particularly relevant to the discussion is Pamela Smith, "In the Workshop of History: Making, Writing, and Meaning," *West 86th: A Journal of Decorative Arts, Design History, and Material Culture* 19 (2012): 4–31.

16. Eamon, *Science and the Secrets of Nature*; William Eamon, *The Professor of Secrets: Mystery, Medicine and Alchemy in Renaissance Italy* (Washington, DC: National Geographic Society, 2010). For a recent study of women, books of secrets, and alchemy, see Meredith Ray, *Daughters of Alchemy: Women and Scientific Culture in Early Modern Italy* (Cambridge, MA: Harvard University Press, 2015).

17. Pamela Smith, "Science on the Move: Recent Trends in the History of Early Modern Science," *Renaissance Quarterly* 62 (2009): 345–75; David Edgerton, *The Shock of the Old: Technology and Global History since 1900* (London: Profile Books, 2006); Francesca Bray, *Technology, Gender and History in Imperial China: Great Transformations Reconsidered* (Abingdon, UK: Routledge, 2013), introduction. See also Francesca Bray, "Gender and Technology," *Annual Review of Anthropology* 36 (2007): 37–53.

18. See, for example, essays in Matteo Valleriani, ed., *The Structures of Practical Knowledge* (Heidelberg: Springer, 2017).

19. Londa Schiebinger, *The Mind Has No Sex? Women in the Origins of Modern Science* (Cambridge: Cambridge University Press, 1989); Margaret Pelling, *Medical Conflicts in Early Modern London: Patronage, Physicians, and Irregular Practitioners, 1550–1640* (Oxford: Clarendon Press, 2003), chap. 6; Susan Broomhall, *Women's Medical Work in Early Modern France* (Manchester: Manchester University Press, 2011); and essays in Sharon Strocchia, ed., "Women and Healthcare in Early Modern Europe," *Renaissance Studies*, special issue, 28, no. 4 (2014): 579–96.

20. Tara Nummedal, "Anna Zieglerin's Alchemical Revelations," in *Secrets and Knowledge in Medicine and Science, 1500–1800*, ed. Elaine Leong and Alisha Rankin (Farnham, UK: Ashgate, 2011), 125–42; Alisha Rankin, *Panaceia's Daughters: Noblewomen as Healers in Early Modern Germany* (Chicago: University of Chicago Press, 2013); Ray, *Daughters of Alchemy*; Sheila Barker, "Christine de Lorraine and Medicine at the Medici Court," in *Medici Women: The Making of a Dynasty in Grand Ducal Tuscany*, ed. Judith C. Brown and Giovanna Benadusi

(Toronto: Centre for Reformation and Renaissance Studies, 2015), 155–81. For women's activities in making natural knowledge, see, for example, Alix Cooper, "Picturing Nature: Gender and the Politics of Natural-Historical Description in Eighteenth-Century Gdańsk/Danzig," *Journal for Eighteenth-Century Studies* 36, no. 4 (2013): 519–29; and Mary Terrall, *Catching Nature in the Act: Réaumur and the Practice of Natural History in the Eighteenth Century* (Chicago: University of Chicago Press, 2013).

21. Christine von Oertzen, Maria Rentezi, and Elizabeth Watkins, eds., "Beyond the Academy: Histories of Gender and Knowledge," *Centaurus* 55, no. 2 (2013): 73–80; Erika Lorraine Milam and Robert A. Nye, eds., "Scientific Masculinities," *Osiris*, special issue, 30 (2015); Donald L. Opitz, Staffan Bergwik, and Brigitte Van Tiggelen, *Domesticity in the Making of Modern Science* (New York: Springer, 2016); Nina E. Lerman, Ruth Oldenziel, and Arwen Mohun, eds., *Gender and Technology: A Reader* (Baltimore: Johns Hopkins University Press, 2003).

22. On "ways of knowing," see John V. Pickstone, *Ways of Knowing: A New History of Science, Technology and Medicine* (Manchester: Manchester University Press, 2000).

23. Harkness, *Jewel House*; Lauren Kassell, *Medicine and Magic in Elizabethan London: Simon Forman, Astrologer, Alchemist, and Physician* (Oxford: Oxford University Press, 2005).

24. Surviving letters suggest that Johanna St. John and John Locke were friends. In the first of the two letters we have, dated 28 April 1693, Johanna remonstrated with Locke for not staying with her at Battersea when he was last in town. Both letters are filled with general news and gossip, but the main point of both was to press Locke for a second opinion in medical matters: Bodleian MS Locke c. 18, fols. 60r–61v.

25. For a similar emphasis on notions of commonality in the early modern German context, see J. Andrew Mendelsohn and Annemarie Kinzelbach, "Common Knowledge: Bodies, Evidence, and Expertise in Early Modern Germany," *Isis* 108, no. 2 (2017): 259–79.

26. On Bacon, see Graham Rees, "An Unpublished Manuscript by Francis Bacon: *Sylva Sylvarum* Drafts and Other Working Notes," *Annals of Science* 38, no. 4 (1981): 377–412; and Doina-Cristina Rusu and Christoph Lüthy, "Extracts from a Paper Laboratory: The Nature of Francis Bacon's *Sylva Sylvarum*," *Intellectual History Review* 27, no. 2 (2017): 171–202. On Draper and Plat, see Harkness, *Jewel House*, chaps. 5 and 6; and Ayesha Mukherjee, "The Secrets of Sir Hugh Platt," in *Secrets and Knowledge*, ed. Leong and Rankin, 69–86.

27. For an overview, see Gianna Pomata and Nancy G. Siraisi, eds., *Historia: Empiricism and Erudition in Early Modern Europe* (Cambridge, MA: MIT Press, 2005).

28. Ursula Klein and Wolfgang Lefèvre, *Materials in Eighteenth-Century Science: A Historical Ontology* (Cambridge, MA: MIT Press, 2007), 23.

29. For historians of science's focus on the philosophical turn, see Harold J. Cook, "The History of Medicine and the Scientific Revolution," *Isis* 102, no. 1 (2011): 102–8.

30. Pamela Smith, *The Body of the Artisan: Art and Experience in the Scientific Revolution* (Chicago: University of Chicago Press, 2004); Pamela Smith, "What Is a Secret? Secrets and Craft Knowledge in Early Modern Europe," in *Secrets and Knowledge*, ed. Leong and Rankin, 47–66. See also Pamela O. Long, *Artisan/Practitioners and the Rise of the New Sciences, 1400–1600* (Corvallis: Oregon State University Press, 2011).

31. Smith, "In the Workshop of History," 26–27, 24.

32. For example, see the essays in Valleriani, *Structures of Practical Knowledge*.

33. For an overview of recent trends in the history of science and the move toward a more expansive definition of science, see Smith, "Science on the Move." For households as spaces of natural inquiry, see Alix Cooper, "Homes and Households," *Early Modern Science*, edited by Katharine Park and Lorraine Daston, vol. 3 of *The Cambridge History of Science* (Cambridge: Cambridge University Press, 2006), 224–37; Alix Cooper, "Picturing Nature"; Lynette Hunter and Sarah Hutton, eds., *Women, Science and Medicine, 1500–1700: Mothers and Sisters of the Royal Society* (Thrupp, Stroud, Gloucestershire, UK: Sutton, 1997); Steven Shapin, "The House of Experiment in Seventeenth-Century England," *Isis* 79, no. 3 (1988): 373–404; Paula Findlen, "Masculine Prerogatives: Gender, Space and Knowledge in the Early Modern Museum," in *The Architecture of Science*, ed. Peter Galison and Emily Thompson (Cambridge, MA: MIT Press, 1999), 29–58; Deborah E. Harkness, "Managing an Experimental Household: The Dees of Mortlake and the Practice of Natural Philosophy," *Isis* 88, no. 2 (1997): 247–62; and, more recently, Terrall, *Catching Nature in the Act*.

34. Thomas Brugis, *The Marrow of Physick* (London, 1640), 86–87. In an inventory made in 1711, Elizabeth Freke lists kettles, skillets, mortar and pestle, colanders, knives, copper and pewter limbecks, a cold still head, a "marrabath head" (bain marie), and much more (BL, MS Additional 45718, fol. 93v or 180). Johanna St. John instructs her steward Thomas Hardyman to return a glass still to a Mrs. Perkins at Lydiard; see this book, chapter 2 and conclusion.

35. See, for example, the essays in Frederic L. Holmes and Trevor H. Levere, eds., *Instruments and Experimentation in the History of Chemistry* (Cambridge, MA: MIT Press, 2000).

36. Steven Shapin, "The Invisible Technician," *American Scientist* 77, no. 6 (1989): 554–63; Shapin, "The House of Experiment."

37. Shapin, "House of Experiment," 378.

38. Harkness, "Managing an Experimental Household"; Findlen, "Masculine Prerogatives"; Terrall, *Catching Nature in the Act*; Terrall, "Masculine Knowledge, the Public Good, and the Scientific Household of Réaumur," *Osiris* 30 (2015): 182–201. For studies focusing on an earlier context, see also Gadi Algazi, "Scholars in

Households: Refiguring the Learned Habitus, 1480–1550," *Science in Context* 16, no. 1–2 (2003): 9–42.

39. Shapin, "Invisible Technician." Additional examples are taken from Terrall, *Catching Nature in the Act*, and Harkness, "Managing an Experimental Household."

40. For an overview of histories of observation, see Lorraine Daston and Elizabeth Lunbeck, *Histories of Scientific Observation* (Chicago: University of Chicago Press, 2011). On the interest in natural particulars across early modern Europe, see Anthony Grafton and Nancy G. Siraisi, *Natural Particulars: Nature and the Disciplines in Renaissance Europe* (Cambridge, MA: MIT Press, 1999).

41. Naomi Tadmor, *Family and Friends in Eighteenth-Century England: Household, Kinship, and Patronage* (Cambridge: Cambridge University Press, 2001).

42. Shapin, "Invisible Technician."

43. I explore this in detail in Elaine Leong, "Collecting Knowledge for the Family: Recipes, Gender and Practical Knowledge in the Early Modern English Household," *Centaurus* 55, no. 2 (2013): 81–103. Alun Withey also acknowledges male interest in recipe collections in his essay "Crossing the Boundaries: Domestic Recipe Collections in Early Modern Wales," in *Reading and Writing Recipe Books, 1550–1800*, ed. Michelle DiMeo and Sara Pennell (Manchester: Manchester University Press, 2013), 179–202.

44. In recent years there has been a flurry of research on English recipe books, and English manuscript recipe books in particular. Pioneering works include Lucinda McCray Beier, *Sufferers and Healers: The Experience of Illness in Seventeenth-Century England* (London: Routledge and Kegan Paul, 1987), chap. 8; Doreen Evenden Nagy, *Popular Medicine in Seventeenth-Century England* (Bowling Green, OH: Bowling Green State University Popular Press, 1988), chap. 5; Jennifer Stine, "Opening Closets: The Discovery of Household Medicine in Early Modern England" (PhD diss., Stanford University, 1996); Lynette Hunter, "Women and Domestic Medicine: Lady Experimenters, 1570–1620," in *Women, Science and Medicine, 1500–1700: Mothers and Sisters of the Royal Society*, ed. Lynette Hunter and Sarah Hutton, 89–107; Lynette Hunter, "Sisters of the Royal Society: The Circle of Katherine Jones, Lady Ranelagh," in *Women, Science and Medicine, 1500–1700: Mothers and Sisters of the Royal Society*, ed. Lynette Hunter and Sarah Hutton (Thrupp, Stroud, Gloucestershire, UK: Sutton, 1997), 89–107, 178–97; and Elaine Leong and Sara Pennell, "Recipe Collections and the Currency of Medical Knowledge in the Early Modern 'Medical Marketplace,'" in *Medicine and the Market in England and Its Colonies c. 1450–c. 1850*, ed. Mark S. R. Jenner and Patrick Wallis (Basingstoke, UK: Palgrave Macmillan, 2007), 133–52. Studies that focus on analyzing recipe collections as a genre of women's writing include Sara Pennell, "Perfecting Practice? Women, Manuscript Recipes and Knowledge in Early Modern England," in *Early Modern Women's Manuscript Writing: Selected Papers from the Trinity/Trent Colloquium*, ed. Victoria Burke and Jonathan Gibson (Aldershot, UK: Ashgate, 2004), 237–58; Laura L. Knoppers, "Opening the Queen's Closet: Henri-

etta Maria, Elizabeth Cromwell, and the Politics of Cookery," *Renaissance Quarterly* 60 (2007): 464–99; Margaret J. M. Ezell, "Domestic Papers: Manuscript Culture and Early Modern Women's Life Writing," in *Genre and Women's Life Writing in Early Modern England*, ed. Michelle M. Dowd and Julie A. Eckerle (Aldershot, UK: Ashgate, 2007), 33–48; Catherine Field, "'Many Hands Hands': Writing the Self in Early Modern Women's Recipe Books," in *Genre and Women's Life Writing*, ed. Dowd and Eckerle, 49–69; Monica Green, *Making Women's Medicine Masculine: The Rise of Male Authority in Pre-Modern Gynecology* (Oxford: Oxford University Press, 2008), 301–10; Elizabeth Spiller, "Introduction," in *Seventeenth-Century English Recipe Books: Cooking, Physic and Chirurgery in the Works of Elizabeth Talbot Grey and Aletheia Talbot Howard: Essential Works for the Study of Early Modern Women*, ser. 3, part 3, vol. 3, edited by Elizabeth Spiller (Aldershot, UK: Ashgate, 2008), ix–li; Elizabeth Spiller, "Recipes for Knowledge: Maker's Knowledge Traditions, Paracelsian Recipes, and the Invention of the Cookbook, 1600–1660," in *Renaissance Food from Rabelais to Shakespeare: Cultural Readings and Cultural Histories*, ed. Joan Fitzpatrick (Aldershot, UK: Ashgate, 2010), 55–72; Elizabeth Spiller, "Printed Recipe Books in Medical, Political, and Scientific Contexts," in *The Oxford Handbook of Literature and the English Revolution*, ed. Laura Knoppers (Oxford: Oxford University Press, 2012), 516–33; Wendy Wall, *Recipes for Thought: Knowledge and Taste in the Early Modern English Kitchen*, Material Texts (Philadelphia: University of Pennsylvania Press, 2016); and Kristine Kowalchuk, *Preserving on Paper: Seventeenth-Century Englishwomen's Receipt Books* (Toronto: University of Toronto Press, Scholarly Publishing Division, 2017).

45. More details are provided in chapter 2. See also Herbert, *Female Alliances*, 102–16.

46. On recipe collections as maps of social networks, see Pennell, "Perfecting Practice?" On the early modern gift economy, see, for example, Natalie Zemon Davis, *The Gift in Sixteenth-Century France* (Madison: University of Wisconsin Press, 2000); Felicity Heal, *The Power of Gifts: Gift Exchange in Early Modern England* (Oxford: Oxford University Press, 2014); Paula Findlen, "The Economy of Scientific Exchange in Early Modern Italy," in *Patronage and Institutions: Science, Technology and Medicine at the European Court, 1500–1750*, ed. Bruce T. Moran (Rochester, NY: Boydell Press, 1991), 5–24; Ilana Krausman Ben-Amos, *The Culture of Giving: Informal Support and Gift-Exchange in Early Modern England* (Cambridge: Cambridge University Press, 2008); and Herbert, *Female Alliances*. On the economy of obligation, see Craig Muldrew, *The Economy of Obligation: The Culture of Credit and Social Relations in Early Modern England* (Basingstoke, UK: Palgrave Macmillan, 1998).

47. Daniel Woolf, *The Social Circulation of the Past: English Historical Culture, 1500–1730* (Oxford: Oxford University Press, 2003).

48. It is likely that this is the manuscript now in the Wellcome Library, cataloged as MS Western 4338. Thus far I have not been able to identify the book of culinary recipes.

49. Wellcome, MS Western 4338, fols. 3r–7r.

50. Ibid., fols. 14r–18v.

51. Ibid., fols. 25r–37r.

52. For the wide range of ailments addressed in such collections, see Leong, "Medical Recipe Collections," 89–97; and Sally Ann Osborn, "The Role of Domestic Knowledge in an Era of Professionalisation: Eighteenth-Century Manuscript Medical Recipe Collections" (PhD diss., University of London, 2015), chaps. 3 and 4.

53. Wellcome, MS Western 4338, fol. 1r.

54. For the various kinds of texts that accompanied recipes in this period, see Leong, "Medical Recipe Collections," 33–37.

55. For an in-depth discussion on the place of inheritance in the transfer of recipe knowledge, see chapter 3. For a discussion on the collective nature of recipe compilation, see Michelle DiMeo, "Authorship and Medical Networks: Reading Attributions in Early Modern Manuscript Recipe Books," in *Reading and Writing Recipe Books, 1550–1800*, ed. Michelle DiMeo and Sara Pennell (Manchester: Manchester University Press, 2013), 25–46; and Leong, "Collecting Knowledge for the Family."

56. For a detailed history of recipes from the ancient world to the thirteenth century, see Tony Hunt, *Popular Medicine in Thirteenth-Century England: Introduction and Texts* (Cambridge: Brewer, 1990), 1–16; and Gianna Pomata, "The Recipe and the Case: Epistemic Genres and the Dynamics of Cognitive Practices," in *Wissenschaftsgeschichte und Geschichte des Wissens im Dialog — Connecting Science and Knowledge*, ed. Kaspar von Greyerz, Silvia Flubacher, and Philipp Senn (Göttingen: V&R Unipress, 2013), 131–54. The literature on recipe studies is vast; see, for example, Laurence Totelin, *Hippocratic Recipes: Oral and Written Transmission of Pharmacological Knowledge in Fifth- and Fourth-Century Greece* (Leiden: Brill, 2009); Michael McVaugh, "The *Experimenta* of Arnald of Villanova," *Journal of Medieval and Renaissance Studies* 1, no. 1 (1971): 107–18; Michael McVaugh, "Two Montpellier Recipe Collections," *Manuscripta* 20, no. 3 (1976): 175–80; Montserrat Cabré, "Women Healers? Household Practices and the Categories of Health Care in Late Medieval Iberia," *Bulletin of the History of Medicine* 82, no. 1 (2008): 18–51; Carla Nappi, "Bolatu's Pharmacy Trade in Early Modern China," *Early Science and Medicine* 14, no. 6 (2009): 737–64; Marta Hanson and Gianna Pomata, "Medicinal Formulas and Experiential Knowledge in the Seventeenth-Century Epistemic Exchange between China and Europe," *Isis* 108, no. 1 (2017): 1–25; and Hilary Marland, "'The Doctor's Shop': The Rise of the Chemist and Druggist in Nineteenth-Century Manufacturing Districts," in *From Physick to Pharmacology: Five Hundred Years of British Drug Retailing*, ed. Louise Hill Curth (Aldershot, UK: Ashgate, 2006), 79–104.

57. See M. L. Cameron, "The Sources of Medical Knowledge in Anglo-Saxon England," *Anglo-Saxon England* 11 (1982): 135–55; M. L. Cameron, "Bald's *Leech-*

book: Its Sources and Their Use in Its Compilation," *Anglo-Saxon England* 12 (1983): 153–82; J. N. Adams and Marilyn Deegan, "Bald's *Leechbook* and the *Physica Plinii*," *Anglo-Saxon England* 21 (1992): 87–114; and Audrey L. Meaney, "Variant Versions of Old English Medical Remedies and the Compilation of Bald's *Leechbook*," *Anglo-Saxon England* 13 (1984): 235–68.

58. See, for example, Rossell Hope Robbins, "Medical Manuscripts in Middle English," *Speculum* 45, no. 3 (1970): 393–415; Warren R. Dawson, ed., *A Leechbook or Collection of Medical Recipes of the Fifteenth Century* (London: Macmillan, 1934); Faye M. Getz, *Healing and Society in Medieval England: A Middle English Translation of the Pharmaceutical Writings of Gilbertus Anglicus* (Madison: University of Wisconsin Press, 1991); Margaret Sinclair Ogden, ed., *The "Liber de Diversis Medicinis" in the Thornton Manuscript (MS Lincoln Cathedral A5.2)* (London: Humphrey Milford, 1938); and Peter Jones, "The *Tabula Medicine*: An Evolving Encyclopaedia," in *English Manuscript Studies, 1100–1700*, vol. 14, *Regional Manuscripts 1200–1700*, ed. A. S. G. Edwards (London: British Library, 2008), 60–85.

59. All the printed recipe collections cited in this introduction were printed in London unless specified otherwise. All dates refer to publication of the first edition.

60. For in-depth publication details, see Leong, "Medical Recipe Collections," chap. 1. For a list of titles, see ibid., appendix B. See also Mary Fissell, "The Marketplace of Print," in *Medicine and the Market in England and Its Colonies, c. 1450–c. 1850*, ed. Mark Jenner and Patrick Wallis (Basingstoke, UK: Palgrave Macmillan, 2007), 116.

61. For a detailed study of *I Secreti*, see Eamon, *Science and the Secrets of Nature*, esp. chap. 4. For studies on Plat's *The Jewell House*, see Mukherjee, "Secrets of Sir Hugh Platt"; and Harkness, *Jewel House*, chap. 6.

62. Frederic L. Holmes, Jürgen Renn, and Hans-Jörg Rheinberger, eds., *Reworking the Bench: Research Notebooks in the History of Science* (Dordrecht: Kluwer, 2003), vii.

63. In that sense, this book differs from cognate projects such as Alisha Rankin's wonderfully rich case studies of Dorothea of Mansfield, Anna of Saxony, and Elizabeth of Rochlitz, which were all framed by hundreds of letters, inventories, accounts, and other paperwork preserved within the archives of the princely courts; see Rankin, *Panaceia's Daughters*, chaps. 3–5.

64. For more details on the parameters of this survey, see Leong, "Medical Recipe Collections," 20–27. This number is not exhaustive, since institutions such as the Folger Shakespeare Library and the Wellcome Library continue to add to their collections. Projects to create lists of such collections are under way; see, for example, "The Manuscript Cookbooks Survey," accessed 22 September 2015, or the "Resources" section of "The Recipe Project," accessed 22 September 2015, http://recipes.hypotheses.org/.

65. For recent overviews, see Leong and Rankin, *Secrets and Knowledge*; and DiMeo and Pennell, *Reading and Writing Recipe Books*. See also the rich informal offerings on the interdisciplinary collaborative research blog *The Recipes Project* (http://recipes.hypotheses.org).

66. For medieval precedents, see above. On recipes and recipe texts in eighteenth-century England, see Gilly Lehmann, *The British Housewife: Cookery Books, Cooking and Society in Eighteenth-Century Britain* (Totnes, Devon, UK: Prospect Books, 2003); and Osborn, "Role of Domestic Knowledge."

67. Harkness, *Jewel House*.

68. Michelle DiMeo and Sara Pennell, "Introduction," in *Reading and Writing Recipe Books*, 9–10.

69. For a recent articulation of the focus on "material texts," see James Daybell, "Introduction," in *The Material Letter in Early Modern England: Manuscript Letters and the Culture and Practices of Letter-Writing, 1512–1635* (London: Palgrave Macmillan, 2012). For a recent overview of early modern marginalia and reader's marks in books, see William H. Sherman, *Used Books: Marking Readers in Renaissance England* (Philadelphia: University of Pennsylvania Press, 2008).

70. Kassell, *Medicine and Magic*; and Harkness, *Jewel House*.

71. See, for example, Sherman, *Used Books*; and Ann Blair, *Too Much to Know: Managing Scholarly Information before the Modern Age* (New Haven, CT: Yale University Press, 2010). On recipes being continually put into action, see, for example, Smith, "In the Workshop of History"; and Pennell, "Perfecting Practice."

Chapter One

1. Archdale Palmer was the son of William and Barbara Palmer, and after attending Sidney Sussex College in Cambridge, he succeeded his father as lord of Wanlip Manor around 1636. Palmer lived with his large family at Wanlip Manor, which is on the river Soar about five miles north of the city of Leicester. It was bought by Archdale's father, William, sometime in the second decade of the seventeenth century. Grant Uden, the editor of Archdale's recipe notebook, notes that though there is evidence that Palmer was a Parliamentarian in the Civil War, there is little evidence that he played a major role in the events of the war. When he died on 6 August 1673, according to the memorial in St. Nicholas Church in Wanlip, he was survived by his wife, Martha, five sons, and two daughters. The recipe collection of Archdale Palmer is now in the Leicestershire County Record Office. It has been edited by Grant Uden, and a facsimile edition that includes a typescript version of the manuscript was printed in 1985. Uden gives a short biography of Palmer in his "Historical Introduction": Archdale Palmer, *The Recipe Book 1659–1672 of Archdale Palmer, Gent. Lord of the Manor of Wanlip in the County of Leicestershire*, ed. B. G. Grant Uden (Wymondham, UK: Sycamore

Press, 1985), ix–xxii. All following references are to Uden's facsimile of the work. The original manuscript has an index at the beginning of the text that was not reproduced in the facsimile.

2. NYPL, Whitney Collection of Cookbooks, Manuscript 9 (hereafter MS Whitney 9). This manuscript notebook has not been paginated, but the recipes are all numbered. Consequently, throughout this chapter I refer to recipe numbers rather than page numbers. From recipe 224 onward, the writer seems to have lapsed back into the hundreds numbering; thus there are two recipes numbered 125, 126, etc. In these cases, the second occurrence of each recipe number is marked as 125a, 126a, etc.

3. MS Whitney 9, recipe 183. The link is confirmed by the numerous contributions from the Shute family, referred to as cousins, and by the frequent references to Somerset towns such as Bruton, Kilmersdon, and Wells. The 1672 visitation records list three families who could be connected with the collection. The numerous mentions of "cousin Shute" confirm that the manuscript must have belonged to one of the children of Philip Bennett and Mary Shute Bennett of Brewham. They are Philip Bennett the younger, who married Anne Strode; Mary, who married John Walter of West Pennard in Somerset; and Martha, who married John Clement of Meere, Wiltshire. See G. D. [George Drewry] Squibb, Edward Bysshe, and the Harleian Society, *The Visitation of Somersett, and the City of Bristol 1672: Made by Sir Edward Bysshe, Knight, Clarenceux King of Arms/Transcribed and Ed. G. D. Squibb* (London: Harleian Society, 1992), 16–17 (hereafter cited as Squibb, *Visitation of Somerset*).

4. The first cited in MS Whitney 9, fol. 82r, is an unnumbered recipe. The recipe likely refers to Francis Perkins, *A New Almanack* (London: Company of Stationers, 1730); Palmer, *Recipe Book*, 97, 148–49. On the intersection of oral and literate cultures, see Adam Fox, *Oral and Literate Culture in England, 1500–1700* (Oxford: Clarendon Press, 2000).

5. This chronological arrangement is common in early modern note taking in general, particularly as an initial gathering of reading notes or other such information. For an overview of learned note taking, see Blair, *Too Much to Know*, chap. 2. Early modern recipe compilers in particular used several methods of organization to ensure easy access to the information they collated. Common strategies include indexes and tables of contents, alphabetical listing, and using a bound notebook from both the front and the back (to separate the medical and culinary content). A few techniques include grouping information by ailment or disease or by the type of medicine and listing the recipes *a capite ad calcem* (from head to foot); some hark back to the medieval origins of the genre. For a more detailed analysis, see Leong, "Medical Recipe Collections," 30–33.

6. R. W. Dunning, ed., "Brewham," in *Bruton, Horethorne and Norton Ferris Hundreds*, vol. 7 of *A History of the County of Somerset* (Oxford: Oxford University Press, 1999), 6–15. According to the 1667 visitation records, there is a Mompesson

family based in Batecombe, and a daughter, Florentia, was included in the family tree; see Squibb, *Visitation of Somerset*, 47.

7. These recipes are dated 24 April 1697 and 23 July 1698. MS Whitney 9, recipes 70, 71, 72, 75, 76.

8. MS Whitney 9, recipes 71, 72.

9. Wellcome, MS Western 160. I was unable to locate biographical information on Anne Brumwich. However, Rhoda Hussey Fairfax (1616/67–86) is the wife of Henry Fairfax's brother Ferdinando. She was the daughter of Thomas Chapman of London and widow of Thomas Hussey of Lincolnshire: Andrew J. Hopper, "Fairfax, Ferdinando, second Lord Fairfax of Cameron (1584–1648)," in *Oxford Dictionary of National Biography*. The book was obviously in Rhoda's possession when she married Thomas Hussey and was brought to the Fairfax family after her marriage to Ferdinando in 1646, two years before his death. The book was then passed on to her daughter Ursula, who married John Cartwright of Aynho in Northamptonshire. The Dorothy Cartwright named must be the daughter of Ursula Fairfax Cartwright: see J. W. Clay, ed., *Dugdale's Visitation of Yorkshire with Additions* (Exeter, 1901), 190. There do not seem to be substantial overlaps between the Brumwich collection and the *Arcana Fairfaxiana*. The additions written by Rhoda Hussey Fairfax seem to refer more to her first husband's family than to that of her second. There are a number of recipes from Lady Hussey, and one was entered as late as 1684, almost forty years after her marriage to Ferdinando Fairfax.

10. Wellcome, MS Western 160, 171, 180. The loose recipe was inserted between pages 170 and 171. I have not been able to find biographical information on Dr. Catlin.

11. Wellcome, MS Western 7113, fol. 3r. Other examples include BL, MS Sloane 1289, which has sections titled "Severall receipts copied out of the booke of Mad. Jones 1681," "Severall receipts copied out of the booke of Madam Porter 1681," and "Severall Receits of my Mother's which she had chiefly in my Lord Berkly's family": fols. 65r, 76r, 80r.

12. MS Whitney 9, recipe 166a.

13. Lisa Smith, "The Relative Duties of a Man: Domestic Medicine in England and France, ca. 1685–1740," *Journal of Family History* 31, no. 3 (2006): 237–56.

14. Andrew J. Hopper, "Fairfax, Henry (1588–1665)," in *Oxford Dictionary of National Biography*.

15. These recipes list only the title and the Latin names and required quantities of the ingredients used: *Arcana Fairfaxiana Manuscripta: A Manuscript Volume of Apothecaries' Lore and Housewifery Nearly Three Centuries Old, Used and Partly Written by the Fairfax Family*, with an introduction by George Weddell (Newcastle-upon-Tyne, UK: Mawson, Swan and Morgan, 1890), 9–58, 75–96.

16. *Arcana Fairfaxiana*, 1, 2, 5–7, 194–98, 8.

17. Ibid., 38.

18. Ibid., 63.

19. Ibid., 65.

20. Ibid., 150.

21. Ibid., 151–53.

22. Sara Pennell, "Perfecting Practice?" 241–42.

23. MS Whitney 9, recipe 130.

24. Ibid., recipe 149.

25. Ibid., recipe 137a.

26. The recipe title reads, "A receipt for my daughter Katherine when so weeke and colourless from Mirs Florence Mompesson April 24th 1697": MS Whitney 9, recipe 70.

27. MS Whitney 9, recipes 189 (recipe collected in 1709), 219, 220 (recipes collected in 1713).

28. An Edmund Trowbridge of Kilmersdon declared himself "disclaimed" in the 1672 visitation of Somerset. It is likely that these are the Trowbridges whom the Bennetts refer to: Squibb, *Visitation of Somersett*. Although no Rowland Cotton is listed in Munk's *Roll*, three men named Rowland Cotton are listed by Foster and by Venn and Venn. The first was admitted as a pensioner at Trinity Hall in April 1641. The second was admitted to Trinity Hall in July 1682, matriculated in 1686, was awarded a BA in 1687, and became the member of Parliament for Newcastle under Lyme. He died 29 April 1753: John Venn and John Archibald Venn, *Alumni Cantabrigienses: A Biographical List of All Known Students, Graduates and Holders of Office at the University of Cambridge, from the Earliest Times to 1900*, vol. 1 (Cambridge: Cambridge University Press, 1922–54), 403. Foster lists a Rowland Cotton who matriculated in Balliol in October 1660 and gained a BA from St. Mary Hall in 1664: Joseph Foster, *Alumni Oxonienses, 1500–1714*, vol. 1 (Oxford: Oxford University Press, 1891), 334.

29. This lies well within the arguments presented by Ronald C. Sawyer in "Patients, Healers, and Disease in the Southeast Midlands, 1597–1634" (PhD diss., University of Wisconsin–Madison, 1986), 182, where he argues that family and friends were often the first resort during bouts of sickness in the early modern period.

30. Whitney, recipes 158a, 159a, 160a.

31. An undated recipe for "pin and web" from a Margerie Land of Brewham follows, and the Bennetts completed this cluster with a recipe for "eye water to turn the humours" from Mrs. Anne Higdon, taken 21 January 1718/19: MS Whitney 9, recipes 161a, 164a.

32. The recipe is recorded as "A Drinke for my wife for melancholy given her 9th April 1662 by pagett": Palmer, *Recipe Book*, 42.

33. Ibid., 113–15.

34. These were written in not by Palmer but rather in a distinct "new" hand: ibid., 93–99.

35. Paul Slack, *The Impact of Plague in Tudor and Stuart England* (Oxford: Oxford University Press, 1985), 30–34.

36. Ibid., 62.

37. Dr. "Micklethwayth" is likely to refer to Sir John Micklethwaite (1612–82), who was practicing medicine in London during the mid-seventeenth century. A fellow of the College of Physicians, Micklethwaite served as censor, treasurer, and president: Willian Birken, "Micklethwaite, Sir John (*bap.* 1612, *d.* 1682)," in *Oxford Dictionary of National Biography.*

38. BL, MS Additional 45198, fol. 5r.

39. Ibid., fol. 95r.

40. Palmer, *Recipe Book*, 100, 25–26.

41. BL, MS Additional 45718, fol. 122r.

42. Palmer, *Recipe Book*, 42–48. For information on Elizabeth Palmer, see ibid., 239.

43. Ibid., 78.

44. Henry Cholmeley's recipes covered a wide range of ailments and are jotted throughout the book. The main section, written in Henry Fairfax's hand, can be found in *Arcana Fairfaxiana*, 138–41.

45. Ibid., 56.

46. Ibid., 146.

47. Wellcome, MS Western 160, 111.

48. Tadmor, *Family and Friends*, chap. 4.

49. Ibid., 150.

50. "My cousin Penrudock" likely refers to Arundele Freke Penrudock's family. "My aunt Tregonell" refers to Anne Freke Tregonwell, who married John Tregonwell of Milton Abbas, Dorset. Her younger sisters Frances and Judith married Sir George Norton and Sir Robert Austen, respectively: Ralph Freke, John Freke, and William Freke, *A Pedigree, or Genealogye, of the Family of the Freke's, Begun by R. Freke, Augmented by J. Freke, Reduced to This Form by W. Freke, July 14th 1707* (London: Typis Medio-Montanis, 1825), 4–5.

51. Tadmor, *Family and Friends*, 165.

52. Ian Archer's study of Samuel Pepys's social milieu is particularly revealing of the various social gatherings involving family and distant kin. Not only did Pepys keep in contact with his relatives, he also felt obligations toward them: Ian Archer, "Social Networks in Restoration London: The Evidence from Samuel Pepys's Diary," in *Communities in Early Modern England*, ed. Alexandra Shepard and Phil Withington (Manchester: Manchester University Press, 2000), 87–89.

53. For additional information on the importance of social visits to early modern English gentry, see Jane Whittle and Elizabeth Griffiths, *Consumption and Gender in the Early Seventeenth-Century Household: The World of Alice Le Strange* (Oxford: Oxford University Press, 2012), 192ff.; and Felicity Heal, *Hospitality in Early Modern England* (Oxford: Clarendon Press, 1990).

54. Palmer, *Recipe Book*, 66–67.

55. "Father Smith" here refers to Archdale's father-in-law, Thomas Smith, a London merchant: ibid., 61. For information on "Father Smith," see 236.

56. Palmer, *Recipe Book*, 8–9.

57. Ibid., 7.

58. Ibid., 130.

59. Ibid., 76.

60. Ibid., 151–52.

61. Mark Goldie, "Tyrell, James (1642–1718)," in *Oxford Dictionary of National Biography*.

62. Wellcome, MS Western 4887, fols. 1v, 13r. The selection from the Hartlib papers falls on fols. 2r–12v. The recipes from Locke's collection are on fols. 13r–39r. Turned upside down and opened from the back, the notebook housed religious tracts written in Latin, thus providing medicine for Tyrrell's body and also his soul.

63. For further information on the Hartlib circle, see George Henry Turnbull, *Samuel Hartlib: A Sketch of His Life and His Relations to J. A. Comenius* (London: Oxford University Press, 1920); George Henry Turnbull, *Hartlib, Dury and Comenius: Gleanings from Hartlib's Papers* (Liverpool: University Press of Liverpool, 1947); Charles Webster, *The Great Instauration: Science, Medicine and Reform, 1626–1660* (London: Duckworth, 1975); and, more recently, Mark Greengrass, Michael Leslie, and Timothy Raylor, eds., *Samuel Hartlib and Universal Reformation: Studies in Intellectual Communication* (Cambridge: Cambridge University Press, 1994). On the *Ephemerides*, see Richard Yeo, *Notebooks, English Virtuosi, and Early Modern Science* (Chicago: University of Chicago Press, 2014), chap. 4.

64. Leigh T. I. Penman, "Omnium Exposita Rapinae: The Afterlives of the Papers of Samuel Hartlib," *Book History* 19, no. 1 (2017): 1–65.

65. Wellcome, MS Western 4887, fols. 4v–6r. These two extracts correspond to the Hartlib Papers 14/5/8A–14/5/9B and 66/8/1A–B.

66. Ibid., fols. 8r–11v. The recipe from Dr. Mackellow is on folio 8r, and the one from Wood is on folio 10v. The recipe from the London merchant is written in a color of ink similar to that in the section taken from Locke's books, suggesting it might have been added in the 1690s.

67. Ibid., fols. 15v (Clark), 19v (Higgins), 16r and 20r (Jep), 25v (Sydenham), and 26r (Boyle).

68. Ibid., fols. 39v–40r. The collection also contains a selection of remedies for horses from Sir Robert Cotton of Combermere, written in a less practiced hand but annotated by Tyrrell himself; ibid., fols. 40v–45v. The Cotton family was also prominent in Cheshire; Sir Robert was the member of Parliament for Cheshire from 1679 to 1681 and 1689 to 1702: S. Handley, "Sir Robert Cotton, 1st Bt. (c. 1635–1712), of Combermere, Cheshire," in *The History of Parliament: The House of Commons, 1690–1715*, ed. D. Hayon, E. Cruickshanks, and S. Handlet (2002), accessed 12 January 2018, http://www.historyofparliamentonline.org/volume/1690 -1715/member/cotton-sir-robert-1635-1712.

69. Frances Norton published her daughter Grace Gethin's *Reliquiae Gethinianae* in her memory after Gethin died at age twenty-one in 1699; thus Freke's copy of these texts must have been made before Grace's death.

70. George Bate (1608–68) was physician to Charles I and Charles II and also to Oliver Cromwell, as well as an active member of the College of Physicians. His *Pharmacopoeia Bateana* was published posthumously in Latin in 1688, with numerous later editions. The Latin version of the text contains Bate's prescriptions as edited by James Shipton, a London-based apothecary, with a selection of the iatrochemist Jonathan Goddard's medicines. Salmon's 1694 translation is taken from the second edition of the Latin work and contains additional commentary and recipes by Salmon himself. The authorship of the recipes is clearly noted in square brackets throughout the text. The second edition (1700) of the text contained even more recipes from Salmon, to the point that there were a similar number of recipes from both authors. The work continued to be published in the eighteenth century, and an edition was issued in 1720. From her reading notes, it seems that Freke consulted the 1700 edition of the work.

71. Moise Charas (1619–1698) was a physician and apothecary in Paris in the mid-seventeenth century. He once gained the title of *l'artiste démonstrateur au Jardin au roi* and was apothecary to the Duke of Orléans: *Dictionnaire de biographie française*, vol. 8, ed. M. Prévost and Roman d'Amat (Paris: Letouzey et Ané, 1959), 464. This pharmacopoeia was one of his three publications. It was first published in 1676 and was translated into English in 1678. Although a second edition of the French version was issued in 1682, only one edition of the English text exists. See Valentine Fernande Goldsmith, *A Short Title Catalogue of French Books, 1601–1700, in the Library of the British Museum* (Folkstone, UK: Dawsons, 1973), 92.

72. George Sweetman, *The History of Wincanton, Somerset, from the Earliest Times to the Year 1903* (London: Henry Williams, 1903), 219–20.

73. MS Whitney 9, recipes 127a (Clarke) and 129a (Maggs).

74. Palmer, *Recipe Book*, 16. I have not been able to find information on the branch of the Palmer family in Aston (perhaps Aston Flamville?). Abraham Wright (1611–1690) was the rector at Oakham in Rutland during this period and was originally from London, so the citation might refer to him: Stephen Wright, "Wright, Abraham, (1611–1690)," in *Oxford Dictionary of National Biography*.

75. Wellcome, MS Western 4338. Willis, Bate, and Colladon contributed numerous recipes including (rather randomly selected), a "wrist medsine for an Ague" (Willis, fol. 4r), an ointment for burns (Bate, 16v), and a recipe for the "Fitts of the mother" (Colladon, fol. 111r). "My Boyles Balsame of sulphire" is on fol. 18r, and Lady Ranelagh's recipe for cough and phlegm is on fol. 100v.

76. The recipe from Mrs. Patrick is for conception: Wellcome, MS Western 4338, fol. 213v.

77. See, for example, Keith Wrightson, "The 'Decline of Neighbourliness' Revisited," in *Local Identities in Late Medieval and Early Modern England*, ed. Norman L. Jones and Daniel Woolf (Basingstoke, UK: Palgrave Macmillan, 2007), 19–49.

78. Anne Stobart, "The Making of Domestic Medicine: Gender, Self-Help and Therapeutic Determination in Household Healthcare in South-West England in the Late Seventeenth Century" (PhD diss., Middlesex University, 2008), 39.

79. Steven King and Alan Weaver, "Lives in Many Hands: The Medical Landscape in Lancashire, 1700–1820," *Medical History* 44, no. 2 (2000): 173–200.

80. Pennell, "Perfecting Practice?" 251–52. Anne Stobart has coined the term "gift medicine" to describe the exchange of medical services ranging from medical knowledge and advice to actual made-up medicines and hard-to-find ingredients to nursing care: Anne Stobart, "Making of Domestic Medicine," 146–53. Janet Theophano discusses the idea of viewing recipes as gifts: *Eat My Words: Reading Women's Lives through the Cookbooks They Wrote* (New York: Palgrave, 2002), 102–3. See also Ben-Amos, *Culture of Giving*; and Craig Muldrew, *Economy of Obligation*.

81. Rankin, *Panaceia's Daughters*, 33–34. See also Findlen, "The Economy of Scientific Exchange."

82. Ray, *Daughters of Alchemy*, 35–45.

83. See Muldrew, *Economy of Obligation*, 121–97; and Bernard Capp, *When Gossips Meet: Women, Family, and Neighbourhood in Early Modern England* (Oxford: Oxford University Press, 2003), 55–68.

84. BL, MS Additional 70006, fol. 222r.

85. For biographical information on Mary Fane, see Stephen Wright, "Fane, Mildmay, second Earl of Westmorland (1602–1666)," in *Oxford Dictionary of National Biography*. Mary Fane is the granddaughter-in-law of Lady Grace Mildmay, a Tudor gentlewoman well known for her medical knowledge. See chapter 3 and Linda Pollock, *With Faith and Physic: The Life of a Tudor Gentlewoman; Lady Grace Mildmay, 1552–1620* (New York: St. Martin's Press, 1995).

86. BL, MS Additional 70113, bundle 3, letter dated 14 January 1650; BL, MS Additional 70006, fol. 207r, letter dated 25 March 1651. For a detailed discussion of this correspondence, see chapter 5.

87. Folger, Bagot Family Papers, MS l.a. 674.

88. Paula Findlen discusses the notion of natural knowledge as a commodity and its role both in seeking and in cementing patronage in "The Economy of Scientific Exchange."

89. Natalie Zemon Davis, "Beyond the Market: Books as Gifts in Sixteenth-Century France," *Transactions of the Royal Historical Society*, 5th ser., 33 (1983): 69–88; Natalie Zemon Davis, *The Gift in Sixteenth-Century France* (Madison: University of Wisconsin Press, 2000); Jason Scott-Warren, *Sir John Harrington and the Book as Gift* (Oxford: Oxford University Press, 2001).

90. Elizabeth Grey, *A Choice Manual of Rare and Select Secrets in Physick and Chyrurgery* (London, 1653), sig. A2r–v.

91. BL, MS Sloane 3842, fol. 1r. For a further discussion on this particular manuscript, see Leong and Pennell, "Recipe Collections and the Currency of Medical Knowledge," 141–43.

92. BL, MS Additional 34722, fols. 1r, 2r.

93. Herbert, *Female Alliances*, chap. 2; Jane Donawerth, "Women's Poetry and the Tudor-Stuart System of Gift Exchange," in *Women, Writing and the Reproduction of Culture in Tudor and Stuart Britain*, ed. Mary E. Burke, Jane Donawerth, Linda L. Dove, and Karen Nelson (Syracuse, NY: Syracuse University Press, 2000), 3–18.

94. Pennell, "Perfecting Practice?"

95. MS Whitney 9, recipes 74 (Goody Hart), 93 (Goody Whittaker), 124 (Goody Creed), 148 (Goodwife Turner), 149 (Goody Collins), 158 (Goodman Portnell), 159 (Goodwife Engrome), 168 (Goodwife Turner), 181 (Goody King of Sturton), 138a (Goody Margaret Bolster), 148a (Goodman Gerhard Wedmoore), and 155a (Goodman Combes).

96. The collection of the Carey family had a remedy for burns from a Goody Bennett: Bodleian, MS Don. e. 11, fol. 40v.

97. David Postles, "The Politics of Address in Early-Modern England," *Journal of Historical Sociology* 18 (2005): 114. Amy Erickson argues that the term "seems to have been both socially and locally specific" in that it is sometimes used to refer to a certain status, for example, a widow: Amy Louise Erickson, *Women and Property in Early Modern England* (Abingdon, UK: Routledge, 1993), 99.

98. Together they contributed eight recipes: Palmer, *Recipe Book*, 22, 66–67, 108, 120–21, 143.

99. Muldrew, *Economy of Obligation*, 150.

100. For a detailed discussion, see Pelling, *Medical Conflicts*, 148–65.

101. For householders as "smart consumers" mixing home-based care with that available commercially, see Leong and Pennell, "Recipe Collections and the Currency of Medical Knowledge," 143–48; and Leong, "Making Medicines."

102. In the case of Dr. Bate, however, it is less clear whether Johanna garnered the remedies from firsthand medical consultations or from his published work *Pharmacopoeia Bateana*. However, the printed work was not published till the late 1680s and thus a little late compared with the likely time frame for Johanna's recipe collecting. Certainly a few of her recipes bear much similarity to those in the *Pharmacopoeia Bateana*, but by no means do all the recipes exactly match those in the printed book. One recipe that bears great resemblance is one for a burn or a scald (Wellcome, MS Western 4338, fol. 16v) that looks like the recipe for an unguentum antipyretium: George Bate, William Salmon, Frederick Hendrick van Hove, and James Shipton, *Pharmacopoeia Bateana* (London: S. Smith and B. Walford, 1670), 689.

103. Wellcome, MS Western 4338, fol. 212v. Willis contributes a significant number of remedies to ease labor or to take during pregnancy, including "A Drink by Dr Willis to correct & dry up sharp humores & prevent miscarrying & good against the whites" and "Dr Willis for Flouding after delivery," fol. 213r.

104. Ibid., fol. 167r.

105. Ibid., fol. 180v.

106. Bodleian, MS Don. e. 11, fols. 52r and 56r.

107. Elizabeth Lane Furdell, "Bate, George (1608–1668)," and Hugh Trevor-Roper, "Mayerne, Sir Theodore Turquet de (1573–1655)," in *Oxford Dictionary of National Biography*. For Mayerne, see also Brian Nance, *Turquet de Mayerne as Baroque Physician: The Art of Medical Portraiture* (Amsterdam: Rodopi, 2001).

108. Bodleian, MS Don. e. 11, fols. 51v–52r, 53v–56r.

109. Ibid., fols. 51v–52r.

110. Ibid., fols. 54r–56r.

111. Lisa Smith, "Relative Duties of a Man."

112. NYPL, English household cookbook, fol. 8v.

113. BL, MS Additional 40696, fols. 162r (letter) and 164r (recipes). The letter was received on 3 April and marked "entered," suggesting that they were copied into a notebook of recipes.

114. Michael Stolberg, "Learning from the Common Folks: Academic Physicians and Medical Lay Culture in the Sixteenth Century," *Social History of Medicine* 27, no. 4 (2014): 649–67; Michael Stolberg, "'You Have No Good Blood in Your Body': Oral Communication in Sixteenth-Century Physicians' Medical Practice," *Medical History* 59, no. 1 (2015): 63–82.

115. MS Whitney 9, recipe 194.

116. There has been a flurry of recent literature on early modern notions of expertise: see, for example, Eric H. Ash, *Power, Knowledge, and Expertise in Elizabethan England* (Baltimore: Johns Hopkins University Press, 2004); Ursula Klein and Emma C. Spary, eds., *Materials and Expertise in Early Modern Europe: Between Market and Laboratory* (Chicago: University of Chicago Press, 2010); Pamela H. Smith and Benjamin Schmidt, eds., *Making Knowledge in Early Modern Europe* (Chicago: University of Chicago Press, 2007); and Lissa Roberts, Simon Schaffer, and Peter Dear, eds., *The Mindful Hand: Inquiry and Invention from the Late Renaissance to Early Industrialization* (Amsterdam: Koninklijke Nederlandse Akademie van Wetenschappen, 2007).

117. Palmer, *Recipe Book*, 152. "Lugborow" likely refers to modern-day Loughborough.

118. A bone spavin is described in the *Oxford English Dictionary* as "a hard bony tumour or excrescence formed at the union of the splint-bone and the shank in a horse's leg, and produced by inflammation of the cartilage uniting those bones; a similar tumour caused by inflammation of the small hock bones." Temple Hall was the residence of Archdale's brother John Palmer: Palmer, *Recipe Book*, 106, 235.

119. Palmer, *Recipe Book*, 19–20.

120. MS Whitney 9, recipes 158a and 158b.

121. On the importance of experiential knowledge in medical recipes, see Rankin, *Panaceia's Daughters*, chap. 1. For the increased importance of empiri-

cism in learned medicine, see Gianna Pomata, "Praxis Historialis: The Uses of Historia in Early Modern Medicine," in *Historia: Empiricism and Erudition in Early Modern Europe*, ed. Gianna Pomata and Nancy G. Siraisi (Cambridge, MA: MIT Press, 2005), 125–37; Gianna Pomata, "Observation Rising: Birth of an Epistemic Genre, ca. 1500–1650," in *Histories of Scientific Observation*, ed. Lorraine Daston and Elizabeth Lunbeck (Chicago: University of Chicago Press, 2011); and Gianna Pomata, "A Word of the Empirics: The Ancient Concept of Observation and Its Recovery in Early Modern Medicine," *Annals of Science* 68 (2011): 1–25.

122. Wrightson, "The 'Decline of Neighbourliness' Revisited"; Phil Withington, "Company and Sociability in Early Modern England," *Social History* 32, no. 3 (2007): 291–307.

123. For a more detailed explanation of this argument, see Leong, "Collecting Knowledge for the Family."

Chapter Two

1. Wiltshire and Swindon History Centre, MS 3430 (papers of the St. John family of Lydiard House); the letters of Sir Walter and Lady Johanna St. John are cataloged as 3430/22. The archive consists of nearly 200 items, of which 150 date to the 1650s, 1660s, and early 1670s. These include a handful of letters (just over a dozen) between Sir Walter and Hardyman, more than 80 between Johanna and Hardyman, various lists (shopping lists for spices, lists of household linens, etc.), menus, and estate accounts. Although most of the letters are undated, Brian Carne, the modern editor of the St. John papers, has used internal evidence to sequence them. This, along with the few dated letters, suggests that the core of the series was written over a five-year period around 1660. A transcription of these letters is available as *Some St. John Family Papers: Transcribed, with Comments, by Canon Brian Carne*, published by the Friends of Lydiard Tregoz (1996). These papers were first published in the annual *Report of the Friends of Lydiard Tregoz* (vols. 27–29). The reports and the papers can be obtained directly from the organization (some reports have been digitized, but these particular volumes are awaiting digitizing): https://www.friendsoflydiardpark.org.uk/. Since the manuscript letters at the Wiltshire and Swindon History Centre have not been foliated, for ease of reference all citations will refer to the Carne edition. I have compared the transcriptions with the original letters, and for the most part these are accurate: Carne, *Some St. John Family Papers*, 69–70.

2. Ibid., 70.

3. For a detailed description, see R. W. Dunning et al., "Parishes: Lydiard Tregoze," in *A History of the County of Wiltshire*, vol. 9, ed. Elizabeth Crittall (London: Victoria County History, 1970), 75–90, accessed 17 June 2015, http://www.british-history.ac.uk/vch/wilts/vol9/pp75-90.

4. Carne, *Some St. John Family Papers*, 49–50.

5. For a recent overview of household management, see Whittle and Griffiths, *Consumption and Gender*, chap. 2; and Steve Hindle, "Below Stairs at Arbury Hall: Sir Richard Newdigate and His Household Staff, c. 1670–1710," *Historical Research* 85, no. 227 (2012): 71–88.

6. By the mid-seventeenth century, English householders could consult several domestic handbooks on housewifery and husbandry. Books such as Gervase Markham's popular manual *The English Hus-wife* and Hannah Woolley's *The Compleat Servant-Maid* offered practical instructions on household tasks, including tips on producing medicines, cookery, gardening, distilling, banquets, wool and cloth dyeing, dairy work, and brewing: Gervase Markham, *The English Housewife* (London: Harison, 1631), contents, sigs. A4r–A6v; and Hannah Woolley, *The Compleat Servant-Maid, or The Young Maidens Tutor* (London, 1677). Woolley's work offers directions for employment as a "waiting woman, House-keeper, Chamber-Maid, Cook-Maid, Under Cook-Maid, Nursery Maid, Dairy-Maid, Laundry-Maid, House-Maid and Scullery-Maid" (title page).

7. The classic study on women and work in early modern England remains Alice Clark's magisterial *Working Life of Women in the Seventeenth Century* (London: Routledge, 1919; and many subsequent editions). Recent studies include Herbert, *Female Alliances*; Jane Whittle, "The House as a Place of Work in Early Modern Rural England," *Home Cultures* 8, no. 2 (2011): 133–50; and Tim Reinke-Williams, *Women, Work and Sociability in Early Modern London* (Basingstoke, UK: Palgrave Macmillan, 2014). For discussions of domestic work as technology, see Bray, *Technology, Gender and History*.

8. For a recent overview of prescriptive literature on domestic labor, see Herbert, *Female Alliances*, 90–102. For a recent discussion on household conduct books, see Harvey, *Little Republic*, chap. 2.

9. For example, Steve Hindle's analysis of the Warwickshire-based Newdigate family reveals the intricate relationship between the master of a rural gentry estate and his various servants: Hindle, "Below Stairs at Arbury Hall." Additionally, Jane Whittle and Elizabeth Griffiths's study of the Le Strange family account books illustrates the wide range of duties taken on by Alice Le Strange and the family's differentiated gendered spending. Their analysis of the accounts offers nuanced readings of the Le Stranges' food production and consumption across seasons and feasts and at different points in their lives: Whittle and Griffiths, *Consumption and Gender*; Jane Whittle, "Enterprising Widows and Active Wives: Women's Unpaid Work in the Household Economy of Early Modern England," *History of the Family* 19, no. 3 (2014): 283–300; Jane Whittle, "Housewives and Servants in Rural England, 1440–1650: Evidence of Women's Work from Probate Documents," *Transactions of the Royal Historical Society*, 6th ser., 15 (2005): 51–74.

10. Carne, *Some St. John Family Papers*, 12, 16, 36, 39, 42–43, 56, 67, 77, 93, 98.

11. Ibid., 42–43.

12. While D. R. Hainsworth has made detailed studies of the letters between squire and estate steward to reveal the rich set of issues involved in managing rural estates from afar, studies of letters between mistresses of elite households and their stewards are scarce: Hainsworth, *Stewards, Lords and People: The Estate Steward and His World in Later Stuart England* (Cambridge: Cambridge University Press, 2008).

13. See, for example, Whittle and Griffiths, *Consumption and Gender*; Hohcheung Mui and Lorna H. Mui, *Shops and Shopkeeping in Eighteenth-Century England* (Kingston, ON: McGill-Queen's University Press, 1989); and Jon Stobart, *Sugar and Spice: Grocers and Groceries in Provincial England, 1650–1830* (Oxford: Oxford University Press, 2013).

14. Carne, *Some St. John Family Papers*, 78–93. In line with studies of dining habits of the period, it is clear that the leftover turkey meat was used in later meals. The menus reveal runs of dishes based on turkey, usually starting with a serving of whole turkey (roasted or boiled), followed by dishes such as cold turkey or turkey pie a couple of meals later. See Carne, *Some St. John Family Papers*, 80–93.

15. A series of daily menus have survived for 15 December 1662 to 11 January 1663. While these perhaps do not represent the family's everyday fare, they show the scale and scope of foods on the table: Carne, *Some St. John Family Papers*, 80–93.

16. Joan Thirsk, *Food in Early Modern England: Phases, Fads, Fashions, 1500–1760* (London: Bloomsbury Academic, 2007), 254; C. Anne Wilson, *Food and Drink in Britain: From the Stone Age to Recent Times* (London: Constable, 1973), 128–31.

17. Instructions for transporting foodstuffs are dotted throughout the letters. In one letter Johanna advises Hardyman that she feels it would be sufficient if he sent up eighteen turkeys and a dozen geese: Carne, *Some St. John Family Papers*, 31. See also, for example, 32 (bacon and pigs), 54 (venison, butter, and turkey eggs), 61 (geese, bacon, turkeys, and rabbits), 63 (pig), and 68 (bacon and turkeys).

18. Ibid., 47–48.

19. Ibid., 29.

20. Woolley, *Compleat Servant-Maid*, 158.

21. Carne, *Some St. John Family Papers*, 32.

22. Ibid., 60.

23. Ibid., 61.

24. Ibid., 62.

25. Ibid., 66.

26. Though Bess was unaware of this practice, it was recommended in contemporary husbandry manuals and seems to have been commonplace in the period. See, for example, the translation of Conrad Heresbach's *Rei rusticate libri quatuor*, which advises feeding them "in like sorte as you doe the Peacocke, or other Poult-

rie; for thei will eate any thyng, and delight in Grasse, Weedes, Gravell, and Sande": Conrad Heresbach, *Foure Bookes of Husbandry* (London, 1578), fol. 167v. Owing to slight inconsistencies between the Carne transcription and the original letter, the transcription here is from the letter itself: Wiltshire and Swindon History Centre, MS 3430/22/1.

27. For brewing, see Elaine Leong, "Brewing Beer and Boiling Water in 1651," in *The Structures of Practical Knowledge*, ed. Matteo Valleriani (Heidelberg: Springer, 2017), 55–75.

28. For the English context, recent studies include Heal, *Power of Gifts*; Ben-Amos, *Culture of Giving*; and Herbert, *Female Alliances*.

29. Heal, *Power of Gifts*, 35–43.

30. Ibid., 36.

31. Heal offers a detailed explanation of the cultural significance of venison: ibid., 40. See also Felicity Heal, "Food Gifts, the Household, and the Politics of Exchange in Early Modern England," *Past and Present* 199 (2008): 58–62.

32. Carne, *Some St. John Family Papers,* 77, 51, 46.

33. In one letter Johanna wrote, "yur Master wil send yu word himself wht he wil have done about the venison he wil alow 3 deer among thos you writ for": ibid., 77.

34. Ibid., 66.

35. Ibid., 56, 60.

36. Ibid., 68, 95.

37. Ibid., 52. Experiments like this are fairly common in the period and are the mainstay of books of secrets. Contemporary with St. John, see, for example, the instructions "How to make Grapes, and other Fruit to have no stone or kernels," in John White's *A Rich Cabinet, with Variety of Inventions* (London, 1689), 37. These are followed by instructions "to make Apples, Pears and other Fruits of several co lours and to give them a dainty tast of Spices": ibid., 38.

38. Ibid., 70.

39. John Gerard, *The Great Herball, or Generall Historie of Plantes* (London, 1597), 852–53.

40. Wellcome, MS Western 3009, 91.

41. Wellcome, MS Western MS/MSL/2, 44.

42. Rebecca Laroche and Steven Turner, "Robert Boyle, Hannah Woolley, and Syrup of Violets," *Notes and Queries* 58 (2011): 390–91.

43. There is a recent flourishing of research on the topic. See, for example, Rebecca Bushnell, *Green Desire: Imagining Early Modern English Gardens* (Ithaca, NY: Cornell University Press, 2003); Leah Knight, *Of Books and Botany in Early Modern England* (Farnham, UK: Ashgate, 2009); and Jennifer Munroe, *Gender and the Garden in Early Modern English Literature* (Aldershot, UK: Ashgate, 2008).

44. Rebecca Bushnell, "Gardener and the Book," in *Didactic Literature in England, 1500–1800: Expertise Construed*, ed. Natasha Glaisyer and Sara Pennell (Aldershot, UK: Ashgate, 2003), 118–36.

45. See, for example, Rebecca Laroche, *Medical Authority and Englishwomen's Herbal Texts, 1550–1650* (Aldershot, UK: Ashgate, 2009).

46. Devon Record Office 1262 M/FC/7. For a discussion of this notebook, see Elaine Leong, "'Herbals She Peruseth': Reading Medicine in Early Modern England," *Renaissance Studies* 28, no. 4 (2014): 556–78.

47. Jennifer Munroe, "Mary Somerset and Colonial Botany: Reading between the Ecofeminist Lines," *Early Modern Studies Journal* 6 (2014): 100–128; Jennifer Munroe, "'My Innocent Diversion of Gardening': Mary Somerset's Plants," *Renaissance Studies* 25, no. 1 (2011): 111–23.

48. For additional examples of gentlewomen taking an interest in their gardens and working in them, see Munroe, *Gender and the Garden*, chap. 1.

49. Carne, *Some St. John Family Papers*, 51.

50. Bushnell, *Green Desire*; and Munroe, *Gender and the Garden*, esp. chap. 1.

51. Carne, *Some St. John Family Papers*, 78.

52. Ibid., 63.

53. Ibid.

54. Ibid., 110.

55. For recent studies exploring discussions of sickness and health in early modern correspondence, see, for example, Weisser, *Ill Composed*; and Newton, *Sick Child*.

56. Carne, *Some St. John Family Papers*, 57.

57. Ibid., 95.

58. For other examples of early modern men offering medical advice and shaping medical encounters for their family members, see, for example, Smith, "Relative Duties of a Man."

59. Carne, *Some St. John Family Papers*, 39.

60. BL, MS Additional 70007, fol. 22r. The Dr. Bathurst mentioned here likely is John Bathurst (d. 1659). He was a member of the College of Physicians and physician to Oliver Cromwell. For more information, see *Oxford Dictionary of National Bibliography*.

61. For the English context, see, for example, Fissell, "Women, Health and Healing"; Leong, "Making Medicines"; and Stobart, *Household Medicine*. For the wider European context, see Rankin, *Panaceia's Daughters* (Germany); and Ray, *Daughters of Alchemy* (Italy).

62. See, for example, Pollock, *With Faith and Physic*.

63. Carne, *Some St. John Family Papers*, 19.

64. Wellcome, MS Western 4338, 30r (citron water using citrons, oranges, and lemons requiring distillation using a limbeck), 174r (recipe for spirit of oranges), 174v (Lady Compton's recipe for spirit of lemons), and 175r ("The Dutches of Lotherdales spirit of Cittrons").

65. Gerard, *Great Herball*, 1465.

66. Wellcome, MS Western 4338, 30r.

67. Carne, *Some St. John Family Papers*, 47–48.

68. Gideon Harvey, *The Family-Physician, and the House-Apothecary* (London: M. Rookes, 1676), A5v.

69. *The Experienced Market Man and Woman, or Profitable Instructions to All Masters and Mistresses of Families, Servants and Others* (Edinburgh: James Watson, 1699), 2.

70. Ibid., 7.

71. For a fuller explanation of the term "smart consumers," see Elaine Leong and Sara Pennell, "Recipe Collections and the Currency of Medical Knowledge."

72. Wellcome, MS Western 4338, fol. 125v.

73. Ibid., fols. 48r, 43r.

74. Carne, *Some St. John Family Papers*, 53.

75. Wellcome, MS Western 4338, fol. 32v.

76. Ibid., fol. 33v.

77. For example, the Lady Childes's ointment to strengthen the sight, an excellent recipe for the gout, or "Oyl of Exiter or my Lord Dennys oyntment for spranes Aches Gout Sciatica Bruse": ibid., fols. 53v, 72v, 122r.

78. Leong, "Making Medicines."

79. The surviving letters do not specify what the boy's illness was. Descriptions of similar necklaces can be found in contemporary herbals such as Gerard's *Great Herball* and John Quincy's *Compleat English Dispensatory* (London, 1718). See Francis Doherty, "The Anodyne Necklace: A Quack Remedy and Its Promotion," *Medical History* 34, no. 3 (1990): 269.

80. Carne, *Some St. John Family Papers*, 63.

81. Ibid., 45.

82. Ibid., 77.

83. Ibid., 53 (red rose water and conserve of red roses) and 74 (candying cowslips).

84. Woolley, *Complete Servant-Maid*, 5–16. See, in particular, the section on "How to Candy all sorts of Flowers as they grow with their Stalk on," 11. Woolley's recipe differs slightly from Johanna's.

85. R. C. Richardson, *Household Servants in Early Modern England* (Manchester: Manchester University Press, 2010), chap. 7.

86. See, for example, Rankin, *Panaceia's Daughters*, chap. 3; and Herbert, *Female Alliances*, 55–62.

87. Frank T. Smallwood, "Lady Johanna St. John as a Medical Practitioner," *Friends of Lydiard Tregoz Report* 6 (1973): 15.

88. Johanna beseeches Hardyman to be sure he or William Franklin is with Sir Walter to give no opportunity to those who want to "vent ther lyes": Carne, *Some St. John Family Papers*, 13.

89. Ibid., 33. At the time one particular letter was written, little Henry St. John had been sent to Lydiard. Johanna gives strict instructions that Henry be lodged with the vicar and his wife and that Hardyman organize a daily delivery of rabbits and see that cocks are sent to the Dewell household regularly and frequently.

90. Ibid., 47.

91. Ibid.

92. Richardson, *Household Servant*; Hindle, "Below Stairs at Arbury Hall"; Hainsworth, *Stewards, Lords and People.*

93. Carne, *Some St. John Family Papers*, 102.

94. Ibid., 97.

95. Ibid., 98.

96. Ibid., 101.

97. Ibid., 99–101.

98. See, for example, Jane O'Hara-May, "Foods or Medicines? A Study in the Relationship between Foodstuffs and Materia Medica from the Sixteenth to the Nineteenth Century," *Transactions of the British Society for the History of Pharmacy* 1 (1971): 61–97; Andrew Wear, *Knowledge and Practice in English Medicine, 1550–1680* (Cambridge: Cambridge University Press, 2000), chap. 4; Sandra Cavallo and Tessa Storey, *Healthy Living in Late Renaissance Italy* (Oxford: Oxford University Press, 2013); and Ken Albala, *Eating Right in the Renaissance* (Berkeley: University of California Press, 2002).

99. Shapin, "Invisible Technician."

Chapter Three

1. James Knowles, "Conway, Edward, second Viscount Conway and second Viscount Killultagh (*bap.* 1594, *d.* 1655)," in *Oxford Dictionary of National Biography.* For a recent comprehensive study of Conway and his literary pursuits, see Daniel Starza Smith, *John Donne and the Conway Papers: Patronage and Manuscript Circulation in the Early Seventeenth Century* (Oxford: Oxford University Press, 2014), esp. chaps. 5 and 6.

2. Conway, it seems, was the one pushing for the correspondence. In a letter early in the series, he asked Harley to send news, and a few letters later Conway entreated him, if it would not be too much trouble, to write every week: BL, MS Additional 70113, bundle 3, letters dated 27 August 1650 and 3 September 1650. Harley was the son of Conway's sister Brilliana and her husband, Sir Robert Harley: Gordon Goodwin, "Harley, Sir Edward (1624–1700)," *Oxford Dictionary of National Biography.*

3. Unfortunately, only part of this series of letters has survived, and these were mostly by Conway. I am grateful to Bernard Capp for bringing the Conway/Harley correspondence to my attention.

4. Recipes also creep into other exchanges within Conway's correspondence. For example, Sir Theodore de Mayerne and Conway shared satirical recipes "for making fat men lean," and Carew Raleigh offers Conway his father's holograph recipe collection: Smith, *John Donne and the Conway Papers*, 117, 112.

5. BL, MS Additional 70113, bundle 3.

6. Ibid.

7. BL, MS Additional 70006, fol. 237r.

8. National Archives, State Papers, Domestic, SP 18/16, fol. 56r–v, letter dated 25 September 1651, Ed. Harley to Viscount Conway at Petworth.

9. While we are certain that Conway worked on at least one recipe book, I have not yet been able to find the notebook. BL, MS Additional 70006, fol. 232r.

10. Ibid., fol. 237r.

11. The letter is dated 25 November 1651: ibid., fol. 247r.

12. Ibid.

13. Daniel Starza Smith, "'La conquest du sang real': Edward, Second Viscount Conway's Quest for Books," in *From Compositor to Collector: Essays on the Book Trade*, ed. Matthew Day and John Hinks (London: British Library/Print Networks, 2012), 191–216; Smith, *John Donne and the Conway Papers*.

14. Elsewhere, examining the brewing of ale and the boiling of water in more detail, I explore the many theoretical frameworks that might have underpinned the "boiling water step": Leong, "Brewing Ale and Boiling Water in 1651," 55–75.

15. The two tracts in question are William Yworth, *Cerevisiarii Comes, or The New and True Art of Brewing* (London: J. Taylor and S. Clement, 1692); and Thomas Tryon, *A New Art of Brewing Beer, Ale and Other Sorts of Liquors* (London: T. Salusbury, 1690).

16. Yworth, *Cerevisiarii Comes*, 16; Tryon, *New Art of Brewing Beer*, 23–24.

17. On the political and social significance of toasting, see Angela McShane, "Material Culture and 'Political Drinking' in Seventeenth-Century England," *Past and Present* 222, suppl. 9 (2014): 247–76.

18. BL, MS Additional 70006, fol. 247r.

19. For a detailed discussion of the value, material or otherwise, of recipes, see Rankin, *Panaceia's Daughters*, chap. 2.

20. Conway's agent Miles Woodshaw wrote in a letter dated 5 September 1650, "I have sent the Iron ffurnace your Lordshipp writ for last weeke, there is another little ffurance and a great still that mr fredrick saith you use to still your roses in": National Archives, State Papers, Domestic, SP 18/11, fol. 13r, letter from Miles Woodshaw to Edward Conway, 5 September 1650.

21. For an articulation of the "invisible" helping hand, in this case in the guise of domestic servants, see Shapin, "Invisible Technician."

22. There is large body of literature on empiricism in the early modern period; see, for example, Pomata and Siraisi, *Historia;* Pomata, "Observation Rising"; and Rankin, *Panaceia's Daughters*, chap. 1.

23. Steven Shapin, *The Social History of Truth: Civility and Science in Seventeenth-Century England* (Chicago: University of Chicago Press, 1994).

24. For a discussion of the importance of making and trying in codifying recipe knowledge, see Rankin, *Panaceia's Daughters*, esp. chap. 1; and Smith, "In the Workshop of History."

25. In viewing Temple's notebooks as "paper technologies" to sort, categorize, and organize information, I participate in a rich conversation. The term "paper technology" was coined by Anke te Heesen in "The Notebook: A Paper-Technology," in *Making Things Public: Atmospheres of Democracy*, ed. Bruno Latour and Peter Weibel (Cambridge, MA: MIT Press, 2005), 582–89. There is now a vast body of literature on this topic. Recent works include monographs like Blair, *Too Much to Know*; and Richard Yeo, *Notebooks, English Virtuosi, and Early Modern Science* (Chicago: University of Chicago Press, 2014). See also special issues of journals: Daniel Rosenberg, ed., "Early Modern Information Overload," *Journal of the History of Ideas* 64 (2003); Richard Yeo, ed., "Note-Taking in Early Modern Europe," *Intellectual History Review* 20, no. 3 (2010); Volker Hess and Andrew Mendelsohn, eds., "Paper Technology in der Frühen Neuzeit," *NTM Zeitschrift für Geschichte der Wissenschaften, Technik und Medizin* 21, no. 1 (2013); and Isabelle Charmantier and Stefan Müller-Wille, eds., "Worlds of Paper," *Early Science and Medicine* 19 (2014). See also Lorraine Daston, "Taking Note(s)," *Isis* 95, no. 3 (2004): 443–48; Daston, "The Sciences of the Archive," *Osiris* 27, no. 1 (2012): 156–87; Michael Stolberg, "Medizinische Loci communes: Formen und Funktionen einer ärztlichen Aufzeichnungspraxis im 16. und 17. Jahrhundert," *NTM Zeitschrift für Geschichte der Wissenschaften, Technik und Medizin* 21, no. 1 (2013): 37–60; Fabian Krämer, "Ulisse Aldrovandi's *Pandechion Epistemonicon* and the Use of Paper Technology in Renaissance Natural History," *Early Science and Medicine* 19 (2014): 398–423; and Fabian Krämer and Helmut Zedelmaier, "Instruments of Invention in Renaissance Europe: The Cases of Conrad Gesner and Ulisse Aldrovandi," *Intellectual History Review* 24, no. 3 (2014): 321–41.

26. Frederic L. Holmes, "Laboratory Notebooks and Investigative Pathways," in *Reworking the Bench: Research Notebooks in the History of Science*, ed. Frederic L. Holmes, Jürgen Renn, and Hans-Jörg Rheinberger (Dordrecht: Kluwer, 2003), 295–308.

27. Sir Peter Temple was the son of John Temple of Stowe, Buckinghamshire, and Dorothy Lee (d. 1625), daughter of Edmund Lee of Stantonbury, Buckinghamshire, and the brother of Sir Thomas Temple (*bap.* 1614, *d.* 1674), the colonial governor of Nova Scotia: John G. Reid, "Temple, Sir Thomas, first Baronet (*bap.* 1614, *d.* 1674)," in *Oxford Dictionary of National Biography.*

28. BL, MS Stowe 1077 (Elinor's gift book), 1078 (intermediary book), and 1079 (vade mecum). Sara Pennell has written evocatively about how different recipe books might serve distinct roles within a household: "Making Livings, Lives and Archives: Tales of Four Eighteenth-Century Recipe Books," in *Reading and*

Writing Recipe Books, ed. Michelle DiMeo and Sara Pennell (Manchester: Manchester University Press, 2013), 225–46.

29. The volume begins with three separate tables of contents. These are titled "A Table of Diseases," "A Particular Table of proved and Experimented Receits," and "A Table beginning at the 400 page of mixt things Belonging to Horses, Husbandry, Cookery, Fishing & other Experiments": BL, MS Stowe 1077, fols. 1r–9r.

30. On the cipher, Temple writes, "I have used this cipher not only to conceale it if it come into an enemys hand but because even ffreinds doe dispise what they know, which is the best excuse for Physitians Bills, & to keepe ignorant persons from applying them (too often) to the Patients prejudice": BL, MS Stowe 1077, fol. 10r–v (10r).

31. Sir Peter Temple was the son of John Temple and Frances Bloomfield Alston. He married Eleanor Tyrell, daughter of Timothy Tyrell of Okely. The collection contains a number of recipes from Sir Edward Tyrell, Lady Elizabeth Tyrell, and Lady Hester Alston. I have not been able to locate Hester Alston, but the Edward Tyrell quoted is probably the uncle of Peter Temple. Tyrell (1573–1656) had two wives named Elizabeth, both of whom died before Temple; thus the recipes could have been donated by either of them.

32. BL, MS Stowe 1078, fol. 45r; and 1077, fol. 35r.

33. BL, MS Stowe 1077, fol. 37r.

34. On the front flyleaf, the book has the ownership note of "Elianor Temple," but it is uncertain whether this refers to Peter's wife or his daughter. Despite the ownership note, Peter played a major role in creating this book and wrote in many of the recipes, the table of contents, and the pagination. Among the recipes are a number collected from his maternal uncle and aunt Sir Edward and Lady Eliza Tyrell. Significantly, however, even after Peter's death (and thus after his daughter Elinor had received her gift book), the Temple women continued to add to this notebook. Both mother and daughter ended up marrying men from the Granville family of nearby Wotton Underwood, and the named contributors and recipe recipients reflect this new extended family network. For example, the book contains a remedy for kidney stones given by Mr. Millins to Mr. Grenville: 1078, fol. 4r.

35. For Francis Bacon's concept of "neat" and "waste" books, see Angus Vine, "Commercial Commonplacing: Francis Bacon, the Waste-Book, and the Ledger," in *Manuscript Miscellanies, 1450–1700*, ed. Richard Beadle, Peter Beal, Colin Burrow, and A. S. G. Edwards (London: British Library, 2011), 197–218.

36. BL, MS Stowe 1078, fol. 25r.

37. This was a remedy for the "falling palatt of the mouth": BL, MS Stowe 1078, fol. 12r; and 1077, fol. 62r.

38. For example, John Aubrey used the same Latin command in his *Naturall Historie*, and the notation also makes frequent appearances in the notebooks of the London merchant Hugh Plat; see Elizabeth Yale, "Marginalia, Commonplaces, and Correspondence: Scribal Exchange in Early Modern Science," *Studies in*

History and Philosophy of Science, Part C: Studies in History and Philosophy of Biological and Biomedical Sciences 42, no. 2 (2011): 197; and Harkness, *Jewel House*, 234–35.

39. BL, MS Stowe 1077, fol. 27r.

40. Nicholas Culpeper, *A Physicall Directory, or A Translation of the London Dispensatory Made by the Colledge of Physicians in London* (London, 1649; and many subsequent editions).

41. BL, MS Stowe 1077, fol. 93r.

42. In his "caution" to Elinor, Temple wrote that even though there are a number of medicines for specific ailments in the collection, "[oftentimes] those universal medicines are the best": BL, MS Stowe 1077, fol. 10r.

43. BL, MS Stowe 1078, fol. 7r. Colonel Lovelace is likely to be Richard Lovelace (1617–1657). See "Lovelace, Richard (1617–1657)," in *Oxford Dictionary of National Biography*.

44. Temple titles this cordial "Sir Richard Napier Cordiall made by old Famous Dr Sandy his Uncle": BL, MS Stowe 1077, fol. 25r. Temple is probably referring to Sir Richard Napier (1607–1676), nephew of the astrological physician Richard Napier (1559–1634), who on the latter's death moved to Great Linford, Buckinghamshire, and took over his uncle's medical practice. One assumes that the "old Famous Dr Sandy his Uncle" refers to Richard Napier the elder: Jonathan Andrews, "Napier, Richard (1559–1634)," in *Oxford Dictionary of National Biography*. For additional information on Richard Napier's life and his medical practice, see Sawyer, "Patients, Healers, and Disease"; and Michael Macdonald, *Mystical Bedlam: Madness, Anxiety and Healing in Seventeenth-Century England* (Cambridge: Cambridge University Press, 1981).

45. BL, MS Stowe 1077, fol. 17r.

46. Ibid., fols. 11v, 44r, 42r.

47. Rankin, *Panacea's Daughters*, 59. On the empiricism in learned medicine, see Pomata, "Praxis Historialis," 125–37; Pomata, "Observation Rising"; Pomata, "Word of the Empirics"; Lauren Kassell, "Casebooks in Early Modern England: Medicine, Astrology, and Written Records," *Bulletin of the History of Medicine* 88, no. 4 (2014): 595–625.

48. BL, MS Stowe 1077, fol. 7r.

49. Claire Jones, "Formula and Formulation: 'Efficacy Phrases' in Medieval English Medical Manuscripts," *Neuphilologische Mitteilungen* 99 (1998): 199–209. For a recent discussion of the use of *probatum est* in manuscript recipe collections, see Wall, *Recipes for Thought*, 212–18.

50. BL, MS Stowe 1077, fol. 35r–v. The Dr. Winston mentioned could refer to Thomas Winston (1575–1655). A fellow of the College of Physicians, Winston acted as a censor from 1622 to 1637 and was made an elect in 1636. He was also professor of physic at Gresham College and physician to Charles I: Ruth Spalding, "Winston, Thomas (c. 1575–1655)," in *Oxford Dictionary of National Biography*.

51. BL, MS Stowe 1077, fol. 23r.

52. Ibid., fol. 18r.

53. Ibid., fol. 52r.

54. Ibid, fol. 52v, and 1079, fol. 54v. The italics indicate phrases that appear only in the gift book.

55. BL, MS Stowe 1077, fol. 53r.

56. Ibid., fol. 41v.

57. Ibid., fol. 45r. Lady Forster may refer to a member of the Forster family who lived locally in Buckinghamshire: John Philipot, William Ryley, William Harvey, William Harry Rylands, and College of Arms (Great Britain), eds., *The Visitation of the County of Buckingham Made in 1634 by John Philipot, Esq.*, Visitation Series 58 (London: Harleian Society, 1909), 55–56, 59.

58. BL, MS Stowe 1077, fol. 24v.

59. For Lady Forster's pomatum, see BL, MS Stowe 1077, fol. 45r; and 1079, fols. 77v–78r.

60. BL, MS Stowe 1077, fol. 62r; 1078, fol. 12r; and 1079, fol. 7r.

61. Adam Smyth, *Autobiography in Early Modern England* (Cambridge: Cambridge University Press, 2010), chaps. 1, 2.

62. For a more detailed discussion of these books, see chapter 5.

63. Unfortunately, I have not been able to find Palmer's other recipe book: Palmer, *Recipe Book*, 138.

64. The book is linked to Godfrey by a recipe labeled "my purg which I take that Mr. Smith the surgon ordered me. Eliz. Godfrey" (with 1686 written in the margin): Wellcome, MS Western 2535, 71.

65. Ibid., 45, 47.

66. Ibid., 3–5.

67. Both these practices can be seen throughout the book, but both are evident in ibid., 11.

68. Ibid., 14, 24, 38.

69. Ibid., 11, 32.

70. Ibid., 18.

71. For a recent nuanced discussion of reader's marks and marginalia in early modern England, see Sherman, *Used Books*.

72. Wellcome, MS Western 3712.

73. Ibid., fols. 117r, 125v.

74. The other manuscript associated with the Okeover family is Wellcome, MS Western 7391. For a detailed history of the two notebooks and additional biographical information on the Okeover family, see Richard Aspin, "Who Was Elizabeth Okeover?" *Medical History* 44, no. 4 (2000): 531–40.

75. Aspin, "Who Was Elizabeth Okeover?" 533–38.

76. Wellcome, MS Western 3712, fols. 185v, 184r.

77. Ibid., fol. 41r.

78. Ibid., fol. 9r. In this period, a lask refers to a looseness in the bowels: "lask, n. 1," *OED Online* (Oxford: Oxford University Press, January 2018), http://www .oed.com/view/Entry/105978?isAdvanced=false&result=1&rskey=eM96Mx&, accessed 4 February 2018.

79. Ibid., fol. 51v.

80. Ibid., fol. 57v.

81. Ibid., fol. 115v.

82. This phrase was written next to two recipes. The first is "Lucatellis his Ballsome: Aunt: L:O: Exelent for many things," and the second is "Blast Salve most exelent." Both these recipes are further marked in the index, with "that I make" written next to "Lucatelus Ballsom" and "tried & is good" next to "ffor a Blast a most exelent oyntment": ibid., fols. 117r, 125v, 185v, 186r.

83. This is written next to "Sr George Hastinges Ballsom" in the index. The actual recipe is marked with a plus sign. Ibid., fols. 126r, 48r.

84. "Rosin, n.," *OED Online*.

85. Perrosin here probably refers to a gum of resin produced from pine trees: "† Perrosin, n.," *OED Online*.

86. Wellcome, MS Western 3712, fol. 114r.

87. Ibid., fol. 192v.

88. Ibid.

89. Ibid., fol. 186r.

90. Ibid., fol. 6v.

91. Ibid., loose leaf at fol. 7r.

92. In Palmer's case, manicules can be found throughout the notebook and were placed against both the headings and sections of the text. The latter practice was used either to emphasize a particular step in the production of the medicament or to highlight a virtue of the drug. See chapters 4 and 5 for more detail on the Fanshawe and Johnson recipe books.

93. These included several recipes for ink and a remedy for "cornes on the feet" by Dr. Clark that Palmer collected in April 1665: Palmer, *Recipe Book*, 73, 83.

94. Mary Grosvenor's recipe book is now BL Sloane 3235. The folio-sized notebook bears the inscription "Mary Grosvenor November 15 1649." The book contains recipes from "my grandmother Cholmeley," further connecting the book with the Cheshire Grosvenor family. Mary's father Sir Richard Grosvenor married Lettice Cholmondeley in 1600. See Richard Cust, "Grosvenor, Sir Richard, first baronet (1585–1645), magistrate and politician," in *Oxford Dictionary of National Biography*.

95. BL, MS Sloane 3235, fols. 2r–8v.

96. This phrase was placed against more than thirty recipes dealing with a variety of ailments: see, for example, Oxford, Exeter College, MS 84, fols. 7r, 15r, 16v, 20v, 21v, 38v.

97. These are sheets inserted within the manuscript itself.

98. Daston, "Sciences of the Archive"; Harkness, *Jewel House*; and Smith, *Body of the Artisan.*

99. Brian W. Ogilvie, *The Science of Describing: Natural History in Renaissance Europe* (Chicago: University of Chicago Press, 2006), chap. 3; Katharine Park, "Natural Particulars: Medical Epistemology, Practice, and the Literature of Healing Springs," in *Natural Particulars: Nature and the Disciplines in Renaissance Europe*, ed. Anthony Grafton and Nancy G. Siraisi (Cambridge, MA: MIT Press, 1999), 347–68.

100. On the relationship of medical records and other kinds of writings such as diaries, registers, and testimonials, see Kassell, "Casebooks."

Chapter Four

1. Paul Seaward, "Dering, Sir Edward, second Baronet (1625–1684)," in *Oxford Dictionary of National Biography.*

2. Huntington, HM MS 41536, fol. 163v.

3. Ibid.

4. Ibid., fol. 30v.

5. Ibid.

6. Ibid., fols. 163v, 34r.

7. The *aqua nephritica* requires milk, maidenhair, goldenrod, pellitory of the wall, nettles, plantain, fennel or sorrel seeds, "knee holly," licorice, and "lignum nephriticum." Readers are instructed first to infuse the herbs, then to steep them overnight and distill the liquid the next morning: Huntington, HM MS 41536, fol. 21r.

8. For an overview of drug testing in learned practices, particularly those of physicians in Europe about 1350 to 1800, see Leong and Rankin, "Testing Drugs and Trying Cures." For detailed studies of how artisans worked through practical knowledge, see Smith, *Bodies of the Artisan*, and Smith, "In the Workshop of History." On practical know-how in general, see essays in Valleriani, *Structures of Practical Knowledge.*

9. These phrases are taken from the notebook of Sir Peter Temple: BL, MS Stowe 1077, fols. 11v, 44r, 48r, 7r, 75r, 23v.

10. Scholars have debated how we might read (and how our historical actors might have read and used) phrases such as *probatum est*, and certainly these phrases cannot always be read as straightforwardly reflecting actual practice. The debate is largely between linguists and centers on medieval English medical recipes. See the differing views of Jones, "Formula and Formulation," and Francisco Alonso-Almeida and Mercedes Cabrera-Abreu, "The Formulation of Promise in Medieval English Medical Recipes: A Relevance-Theoretic Approach," *Neophilologus* 86, no. 1 (2002): 137–54.

11. Folger, MS v.a. 388; Glasgow, MS Ferguson 43.

12. Wellcome, MS Western 4338, fols. 91r, 101v, 199r.

13. Ibid., fol. 86r.

14. Ibid., fol. 180r.

15. Ibid., fols. 46r, 139v.

16. BL, MS Stowe 1077, fol. 7r.

17. Wellcome, MS Western 184a, fol. 41v.

18. Charles B. Schmitt, "Experience and Experiment: A Comparison of Zab-arella's View with Galileo's in *De Motu*," *Studies in the Renaissance* 16 (1969): 80–138; Jole Agrimi and Chiara Crisciani, "Per una ricerca su *experimentum-experimenta*: Riflessione epistemologica e tradizione medica (secoli XIII–XV)," in *Presenza del lessico greco et latino nelle linque contemporanee*, ed. Pietro Janni and Innocenzo Mazzini (Macerata, Italy: Facoltà di Lettere e Filosofia, Università degli Studi di Macerata, 1990), 9–49; Peter Dear, "The Meanings of Experience," in *Early Modern Science*, ed. Katharine Park and Lorraine Daston, vol. 3 of *The Cambridge History of Science* (Cambridge: Cambridge University Press, 2006), 106–31. See also Leong and Rankin, "Testing Drugs and Trying Cures," 157–82; and Evan R. Ragland, "'Making Trials' in Sixteenth- and Early Seventeenth-Century European Academic Medicine," *Isis* 108, no. 3 (2017): 503–28.

19. For discussion of this kind of adjustment in alchemical recipes, see Jenni-fer M. Rampling, "Transmuting Sericon: Alchemy as 'Practical Exegesis' in Early Modern England," *Osiris* 29 (2014): 19–34. For a discussion of cognate practices in the workshops of apothecaries in northern Italy, see Valentina Pugliano, "Phar-macy, Testing and the Language of Truth in Renaissance Italy," *Bulletin of the His-tory of Medicine* 91, no. 2 (2017): 233–73.

20. Huntington, HM MS 41536, fol. 168v.

21. Ibid., fol. 166v.

22. See essays in Leong and Rankin, "Testing Drugs and Trying Cures," particu-larly Leong and Rankin, "Testing Drugs and Trying Cures"; and Alisha Rankin, "On Anecdote and Antidotes: Poison Trials in Sixteenth-Century Europe," *Bulle-tin of the History of Medicine* 91, no. 2 (2017): 280–84. On the practices of Caterina Sforza, see Ray, *Daughters of Alchemy*, chap. 1, esp. 23.

23. On repetition and replication in early modern experiments, see Jutta Schickore, "Trying Again and Again: Multiple Repetitions in Early Modern Re-ports of Experiments on Snake Bites," *Early Science and Medicine* 15 (2010): 567–617; and Jutta Schickore, "The Significance of Re-doing Experiments: A Contribu-tion to Historically Informed Methodology," *Erkenntnis* 75, no. 3 (2011): 325–47.

24. Stephen Pender, "Examples and Experience: On the Uncertainty of Medi-cine," *British Journal for the History of Science* 39, no. 140, pt. 1 (2006): 1–28.

25. For a detailed discussion of Glyd's book and the writing of family history, see chapter 5.

26. Heinz Otto Sibum, "Reworking the Mechanical Value of Heat: Instruments of Precision and Gestures of Accuracy in Early Victorian England," *Studies in the*

History and Philosophy of Science, part A 26, no. 1 (1995): 73–106, and James Sumner, *Brewing Science, Technology and Print, 1700–1880* (London: Routledge, 2013).

27. BL, MS Additional 45196, fol. 32r.

28. Ibid., fol. 55v.

29. Wellcome, MS Western 3082, fol. 68r.

30. Quid pro quo texts have a long tradition; see, for example, Alain Touwaide, "Quid pro Quo: Revisiting the Practice of Substitution in Ancient Pharmacy," in *Herbs and Healing, from the Ancient Mediterranean through the Medieval West*, ed. Ann Van Arsdall and Timothy Graham (Farnham, UK: Ashgate, 2012), 19–61. For contemporary discussions of this, see, for example, *The Country-Man's Apothecary, or A Rule by Which Country Men Safely Walk in Taking Physick; Not Unusefull for Cities. A Treatise Shewing What Herbe, Plant, Root, Seed or Mineral, May be Used in Physicke in the Room of That Which Is Wanting* (London, 1649). This is a translation of Guillaume Rondelet's (1507–1566) *Tractatus de Succedaneis* (Basel, 1587).

31. BL, MS Additional 45196, fol. 54v. Unset leeks are very young leeks that have not yet been planted out.

32. Ibid., fol. 49v.

33. Oxford, Bodleian Library, MS Eng. Misc. d. 436, fol. 179r.

34. BL, MS Additional 45196, fol. 47r.

35. Wellcome, MS Western 7113, fol. 126v.

36. BL, MS Additional 45196, fol. 67v.

37. BL, MS Stowe 1079, fol. 1v.

38. Ibid., fol. 81r.

39. BL, MS Stowe 1078, fol. 25r.

40. BL, MS Stowe 1077, fol. 33r.

41. On the early modern interest in materials, see Ursula Klein and Emma C. Spary, eds., *Materials and Expertise in Early Modern Europe: Between Market and Laboratory* (Chicago: University of Chicago Press, 2010).

42. Wellcome, MS Western 2535, 81.

43. On notions of uncertainty in early modern medicine, see Pender, "Examples and Experience."

44. Smith, "In the Workshop of History." See also Pamela H. Smith, Amy R. W. Meyers, and Harold J. Cook, eds., *Ways of Making and Knowing: The Material Culture of Empirical Knowledge*, Bard Graduate Center Cultural Histories of the Material World (Ann Arbor: University of Michigan Press, 2014). On the difficulty of working with early modern instruments in experiments, see Simon Schaffer, "Glass Works: Newton's Prisms and the Uses of Experiment," in *The Uses of Experiment: Studies in the Natural Sciences*, ed. David Gooding, Trevor Pinch, and Simon Schaffer (Cambridge: Cambridge University Press, 1989), 67–104.

45. Wellcome, MS Western 2535, 5.

46. The herbs are star thistle, marigold leaves, chickweed, alehouse, the inner green rind of elder bark, and herb of grace: BL, MS Additional 45196, fol. 73v.

47. Wellcome, MS Western 3082, fol. 65r.

48. Ibid., fol. 65v.

49. BL, MS Stowe 1077, fol. 13r.

50. BL, MS Stowe 1079, fols. 1v.

51. BL, MS Stowe 1077, fol. 31r; and 1078, fol. 6r.

52. BL, MS Stowe 1078, fol. 6r–v; and 1077, fol. 31r.

53. BL, MS Stowe 1079, fols. 56v–60r; and 1078, fols. 14r–15r.

54. BL, MS Stowe 1077, fols. 98r–101r.

55. Ibid., fol. 98v.

56. Ibid., fols. 100v, 99v.

57. Ibid., fol. 100v.

58. There is a sizable body of work dealing with authorship in compiled texts, especially in poetic miscellanies: see, for example, Marcy North, *The Anonymous Renaissance: Cultures of Discretion in Tudor-Stuart England* (Chicago: University of Chicago Press, 2003); Margaret Ezell, *Social Authorship and the Advent of Print* (Baltimore: Johns Hopkins University Press, 2003); Harold Love, *Scribal Publication in Seventeenth-Century England* (Oxford: Oxford University Press, 1993), reprinted as *The Culture and Commerce of Texts: Scribal Publication in Seventeenth-Century England* (Amherst: University of Massachusetts Press, 1998); Henry R. Woudhuysen, *Sir Philip Sidney and the Circulation of Manuscripts, 1558–1640* (Oxford: Oxford University Press, 1996); and Arthur Marotti, *Manuscript, Print and the English Renaissance Lyric* (Ithaca, NY: Cornell University Press, 1995).

59. Michael McVaugh, "The 'Experience-Based Medicine' of the Thirteenth Century," *Early Science and Medicine* 14, no. 1 (2009): 105–30.

60. Leong and Rankin, "Testing Drugs and Trying Cures."

61. This has been much commented on: see, for example, Steven Shapin and Simon Shaffer, *Leviathan and the Air-Pump: Hobbes, Boyle, and the Experimental Life* (Princeton, NJ: Princeton University Press, 1985); and Peter Dear, "*Totius in Verba*: Rhetoric and Authority in the Early Royal Society," *Isis* 76, no. 2 (1985): 144–61.

62. Klein and Lefèvre, *Materials in Eighteenth-Century Science*, 21–23.

Chapter Five

1. Burford Butcher, "The Brockman Papers," *Archaeologia Cantiana* 43 (1931): 281–83.

2. Edward Hasted, "Parishes: Newington," in *The History and Topographical Survey of the County of Kent* (Canterbury: W. Bristow, 1799), 8:197–210, accessed 17 May 2015, http://www.british-history.ac.uk/survey-kent/vol6/pp40-67.

3. H. I. Bell, "The Brockman Charters," *British Museum Quarterly* 6, no. 3 (1931): 75.

4. For additional information on Ann Bunce Brockman, see Kate Aughterson, "Brockman, Ann, Lady Brockman (*d.* 1660)," in *Oxford Dictionary of National Biography.* See also Giles Drake-Brockman, "Sir William and Lady Ann Brockman of Beachborough, Newington by Hythe: A Royalist Family's Experience of the Civil War," *Archaeologia Cantiana* 132 (2012): 21–41. Lady Brockman is mentioned in relation to Betty's burns as someone else who uses an oil-based remedy: Huntington Library, HM MS 41536, fol. 31v.

5. Michelle DiMeo has recently written sensitively about the books of Anne and Elizabeth Brockman. See DiMeo, "Authorship and Medical Networks."

6. Margaret Ezell has written extensively about the recipe book of Anne Glyd: see Ezell, "Domestic Papers," 33–48; and Margaret Ezell, "Invisible Books," in *Producing the Eighteenth-Century Book: Writers and Publishers in England, 1650–1800*, ed. Pat Rogers and Laura Runge (Newark: University of Delaware Press, 2009), 53–69.

7. These twenty-eight items were separately donated by Phyllis Brockman in 1938 and are now cataloged as additional manuscripts 45193–45220. Additional manuscript 45193 is an anonymous sixteenth- and seventeenth-century commonplace book with drawings depicting well-known engravings and genealogical notes on the Stoughton and Drake families; see below for further discussion. Richard Glyd's instructions for the disposal of his lands and properties are now cataloged as additional manuscript 45195. The French grammar and general notebook of James Brockman constitute additional manuscript 45120. The household accounts of Anne Glyd Brockman dating from 1700 to 1724 are now cataloged as additional manuscripts 45208–45210.

8. See also Janet Theophano, *Eat My Words: Reading Women's Lives through the Cookbooks They Wrote* (New York: Palgrave, 2002); and DiMeo, "Authorship and Medical Networks."

9. On notions of family strategies, particularly in early modern France, see, for example, Natalie Zemon Davis, "Ghosts, Kin, and Progeny: Some Features of Family Life in Early Modern France," *Daedalus* 106, no. 2 (1977): 87–114; Joanna Milstein, *The Gondi: Family Strategy and Survival in Early Modern France* (Aldershot, UK: Ashgate, 2014); and Joanne Baker, "Female Monasticism and Family Strategy: The Guises and Saint Pierre de Reims," *Sixteenth Century Journal* 28 (1997): 1091–108.

10. Daniel Woolf, *The Social Circulation of the Past: English Historical Culture, 1500–1730* (Oxford: Oxford University Press, 2003); Katharine Hodgkin, "Women, Memory and Family History in Seventeenth-Century England," in *Memory before Modernity: Practices of Memory in Early Modern Europe*, ed. Erika Kuijpers et al. (Leiden: Brill, 2013), 297–314.

11. For an overview of recent work in the field, see Elizabeth Yale, "The History of Archives: The State of the Discipline," *Book History* 18, no. 1 (2015): 332–59;

and Alexandra Walsham, "The Social History of the Archive: Record-Keeping in Early Modern Europe," *Past and Present* 230, suppl. 11 (January 1, 2016): 9–48. For a discussion of the history of archives, particularly in relation to the history of science, see Michael Hunter, ed., *Archives of the Scientific Revolution: The Formation and Exchange of Ideas in Seventeenth-Century Europe* (Woodbridge, UK: Boydell Press, 1998).

12. Kathryn Burns, *Into the Archive: Writing and Power in Colonial Peru* (Durham, NC: Duke University Press, 2010), 124.

13. Elizabeth Yale, *Sociable Knowledge: Natural History and the Nation in Early Modern Britain* (Philadelphia: University of Pennsylvania Press, 2016), chap. 6.

14. Field and Theophano, *Eat My Words.*

15. Wellcome, MS Western 184a, fol. 2v.

16. Stine, "Opening Closets," 144. The manuscript has an inscription linking it to both Anne and Alethea Talbot Howard. Charles Howard was the brother of Henry Frederick Howard, fifteenth Earl of Arundel (1608–1652), and was involved in the early Royal Society. However, by the time this manuscript was copied and presented to his niece, Charles was no longer regularly attending the meetings but was devoting his time to other pursuits: see James McDermott, "Howard, Charles, second Baron Howard of Effingham and first Earl of Nottingham (1536–1624)," in *Oxford Dictionary of National Biography.*

17. BL, MS Sloane 3235, 1r (Grosvenor's ownership note), 26v.

18. Wellcome, MS Western 3082, fol. 27r. The Johnson family book was probably begun by Elizabeth Oldfield Philips Johnson in 1694 during her first marriage to Andrew Philips of Leominster. Like the brides mentioned above, she brought the book with her to Spalding when she married Maurice Johnson. For a detailed family tree, see Everard Green, "Pedigree of Johnson of Ayscough-Fee Hall, Spalding, Co. Lincoln," *Genealogist* 1 (1877): 110.

19. David Boyd Haycock, "Johnson, Maurice (1688–1755)," in *Oxford Dictionary of National Biography.*

20. The Johnson family book is well annotated. The first section of the manuscript contains miscellaneous material including newspaper cuttings dating to the first quarter of the nineteenth century: see, for example, Wellcome, MS Western 3082, fols. 3r (a recipe to destroy moths, dated 1827), 6r (information on growing strawberries, dated 1828), and 27v (notes from various newspapers and a clipping of a recipe for "Otto of Roses").

21. See, for example, Yale, *Sociable Knowledge*, chap. 6.

22. Smyth, *Autobiography*, 151.

23. BL, MS Additional 45196, fol. 11v.

24. Ibid., fol. 75v.

25. Jones, "Formula and Formulation."

26. BL, MS Additional 45196, fol. 81v. The same changes in notation can be seen under the recipe for a "cramp," fol. 79v. See chapter 4 for a detailed analysis of issues of testing, trying, and practice.

27. Ibid., fol. 66r.

28. BL, MS Additional 45198, fols. 7, 38.

29. Ibid., fol. 23v.

30. BL, MS Additional 45196, fols. 79v, 58r, 46v.

31. BL, MS Additional 45198, fol. 17r.

32. Ibid., fol. 34r.

33. Lisa Smith, "Sir Hans Sloane, Friend and Physician to the Family," in *From Books to Bezoars: Sir Hans Sloane, Collector of an Age*, ed. Alison Walker, Michael Hunter, and Arthur MacGregor (London: British Library Publications, 2012), 51.

34. Wellcome, MS Western 762, fol. 293r.

35. BL, MS Additional 45718, fol. 122v.

36. Sara Pennell first suggested reading recipe books as maps of family networks and alliances: "Perfecting Practice," 243.

37. Yale, "Marginalia," 197. For a recent study on sociability and the *album amicorum*, see Bronwen Wilson, "Social Networking: The *Album Amicorum* and Early Modern Public-Making," in *Beyond the Public Sphere: Opinions, Publics, Spaces in Early Modern Europe*, ed. Massimo Rospocher (Bologna: Società editrice il Mulino, 2012), 205–23.

38. Woolf, *Social Circulation of the Past*.

39. Cassandra Willughby, Duchess of Chandos, created several manuscripts on her family's history; see her entry in the *Catalogue of English Literary Manuscripts*, http://www.cclm-ms.org.uk/authors/willoughbycassandraduchessofchandos .html. In particular, two volumes titled "An Account of the Willughby's of Wollaton" are now in the Nottingham University Library, cataloged as Mi LM 26 and 27. For more details on the Willughby family history and domestic papers, see Dorothy Johnston, "The Life and Domestic Context of Francis Willughby," in *Virtuoso by Nature: The Scientific Worlds of Francis Willughby FRS (1635–1672)*, ed. Tim Birkhead (Leiden: Brill, 2016), 1–43.

40. Bodleian MS Top Oxon. c. 757. The volume has inscribed on the first folio, "sum liber edmundi ex dono christoferi napperi." The Nappers or Napiers were a prominent Oxfordshire family from Holywell. Both Edmund (1579–1654) and Christopher were sons of William Napier (bur. 1621/22) and Elizabeth Powell Napier (bur. 1584). Edmund Napier's son George (bap. 1619–1671) married Margaret Arden of Kertlington. George and Margaret's daughter Margaret married Henry Nevill of Holt in Leicestershire: G. D. [George Drewry] Squibb, Edward Bysshe, and the Harleian Society, *The Visitation of Oxfordshire 1669 and 1675: Made by Sir Edward Bysshe, Knight, Clarenceux King of Arms/Transcribed and Ed. G. D. Squibb* (London: Harleian Society, 1993), 25–26.

41. Bodleian, MS Top Oxon. c. 757, unfoliated front flyleaf iii verso.

42. Ibid., unfoliated front flyleaf ii verso.

43. Ibid., fols. 1r, 9r, 10r, and more (leases), 2r (rents due), 164v (survey of Margaret and Francis Napier's estate in November 1672), 156r–159v, a list of books (undated but likely from the 1720s), and copies of letters.

44. Ibid., fols. 2v, 6r, 36r, 37r–v.

45. On family history and family strategy, see Davis, "Ghosts, Kin, and Progeny." For an overview on the *libri di famiglia*, see Angelo Cicchetti and Raul Mordenti, *I libri di famiglia in Italia*, vol. 1, *Filologia e storiografia letteraria* (Rome: Edizioni di Storia e Letteratura, 1985). For parallel consideration of the Italian and English family history writing traditions, see Eric Ketelaar, "Muniments and Monuments: The Dawn of Archives as Cultural Patrimony," *Archival Science* 7, no. 4 (2007): 343–57; and Ketelaar, "The Genealogical Gaze: Family Identities and Family Archives in the Fourteenth to Seventeenth Centuries," *Libraries and the Cultural Record* 44, no. 1 (2009): 9–28.

46. Davis, "Ghosts, Kin, and Progeny," 97.

47. Woolf, *Social Circulation of the Past*, 116–17; Hodgkin, "Women, Memory and Family History."

48. Hodgkin, "Women, Memory and Family History," 299–300.

49. See Whittle and Griffith, *Consumption and Gender*, 28, for the case of the Le Strange account books. In this instance the husband, Hamon, was initially responsible for the book, but the wife, Alice, took over a few years later and remained the main "accountant" for the family.

50. For example, within these works: te Heesen, "Notebook"; Blair, *Too Much to Know*; Yeo, *Notebooks, English Virtuosi and Early Modern Science*; Hess and Mendelsohn, "Paper Technology"; Rosenberg, "Early Modern Information Overload"; Charmantier and Müller-Wille, "Worlds of Paper"; and Yeo, "Note-Taking in Early Modern Europe."

51. For recent overview, see the rich essays in Walsham, "Social History of the Archive." See also Jacob Soll, *The Information Master: Jean-Baptiste Colbert's Secret State Intelligence System: Cultures of Knowledge in the Early Modern World* (Ann Arbor: University of Michigan Press, 2009). On accounting and life writing see Smyth, *Autobiography in Early Modern England*; and Jason Scott-Warren, "Early Modern Bookkeeping and Life-Writing Revisited: Accounting for Richard Stonley," *Past and Present* 230, suppl. 11 (2016): 151–70.

52. Victoria Burke, "Ann Bowyer's Commonplace Book (Bodleian Ashmole MS 51): Reading and Writing among the 'Middling Sort,'" *Early Modern Literary Studies* 6, no. 3 (2001): 1–28; Victoria Burke, "Recent Studies in Commonplace Books," *English Literary Renaissance* 43, no. 1 (2013): 153–77; Ezell, "Domestic Papers," 33–63.

53. Wellcome, Western MSS 7997–8002, 8680.

54. Wellcome, MS Western 7998, front cover; and MS Western 7999.

55. Wellcome, Western MS 7998, inside cover. We might also note that the bookplate of the Wellcome Library, with its own naming system, obscures both the Faussett family ownership notes.

56. Wellcome, MS Western 761-2.

57. Wellcome, MS Western 761, fol. 5r; and 762, fol. 5r.

58. Jennifer Heller, *The Mother's Legacy in Early Modern England* (Farnham, UK: Ashgate, 2011).

59. Smyth, *Autobiography*, 149–54. See also Blair, *Too Much to Know*, 86.

60. Folger Shakespeare Library, MS v.a. 430, pastedown on front cover. The notebook was compiled between 1640 and 1750. Mary Granville was the daughter of Sir Martin Wescombe, who was the English consul in Cadiz, Spain, from 1663 to 1688. Anne Granville D'Ewes married John D'Ewes of Wellesbourne, Warwickshire, and resided in Bradley, Worcester: Catherine Field, "Kitchen Science and the Public/Private Role of the Housewife Practitioner" (paper presented at the Women on the Verge of Science seminar at the Folger Institute, May 14, 2003), 4.

61. BL, MS Stowe 1077. This is part of a set of three recipe books, Stowe MSS 1077–79. The books are fully explored in chapters 3 and 4.

62. BL, MS Stowe 1077, fol. 10r.

63. Bodleian, MS Tanner 397, fol. 1r.

64. Ibid., fols. 62r–131v (medical recipes), 33r–61r (culinary recipes), 169v–173v (veterinary recipes), 4r–v (weights and measurements), 5r–v (Giovanni da Vigo on pain), 6r–8v (glossary), 200v–194v (treatise of urine), 205v–1v (extract of Galen's *De Alimentorum Facultatibus*), and 209v–7v (section for preserving health). The book was turned upside down and information was entered from the back of the volume, hence some folio numbers are reversed. The section from da Vigo's *Most Excellent Workes of Chirurgerye* can found on fol. ccv verso in the 1543 edition.

65. Ibid., fols. 9r–10r.

66. Ibid., fols. 187v–73r (list of mayors and sheriffs) and 193v–88r (list of high sheriffs). Here Bourne does not appear to be attempting to construct a local history of Norwich or any such similarly ambitious project. Recording local historical and genealogical information was common. For a detailed discussion of early modern historical culture, see Woolf, *Social Circulation of the Past*. Another example of pairing medical and historical information can be seen in the 1598 edition and many subsequent editions of *The Pronostycacyon for Euer of Erra Pater*.

67. BL, MS Additional 45196, fols. 82r–84r.

68. For example, further genealogical information can be found in BL, MS Additional 45193, containing information on the Stoughton family of Stoughton Place. This information has been annotated by a number of hands, including one that resembles Anne Glyd's.

69. Wellcome, MS Western 4683, fol. 1v.

70. Ibid., fol. 180r.

71. Bodleian, MS Eng. Misc. d. 436, fols. 194v–195r.

72. Though not discussed in detail, other examples of this practice include the book associated with the Grosvenor family that lists the dates of birth of Sidney (16 January 1650), Anne (16 December 1652), Thomas (20 November 1655), Robert (15 April 1657), John (31 January 1658), and Roger (5 December 1661): BL, MS Sloane 2266, fols. 1v–2r. Sara Pennell also discusses it in the case of Hannah Bisaker in "Making Livings, Lives and Archives," 229.

73. Woolf, *Social Circulation of the Past*, 116–17.

74. Sherman, *Used Books*, 58 and chap. 3. For the inclusion of family history in almanacs, see Smyth, *Autobiography*, 20.

75. Stine, "Opening Closets," 111; Pennell, "Perfecting Practice?" 240–41.

76. Smyth, *Autobiography*, 11 and chap. 2.

Chapter Six

1. For a biography of Stuart, see Donald W. Foster, "Stuart, Frances, Duchess of Lennox and Richmond [other married name Frances Seymour, Countess of Hertford] (1578–1639)," in *Oxford Dictionary of National Biography*. The portrait appears to be a copy of Francis Delaram's (1589/90–1627) engraved portrait of Stuart taken from John Smith's (1580–1631) *The Generall Historie of Virginia, New-England, and the Summer Isles* (London: Printed by I[ohn] D[awson] and I[ohn] H[aviland] for Michael Sparkes, 1626). Two slightly different portraits of Frances Stuart were included in the *Generall Historie*. A copy of the portrait that clearly "inspired" Stafford and his book-producing team is at the Beinecke Library at Yale University, call number Z40 025.

2. Owen Wood, *Choice and Profitable Secrets both Physical and Chirurgical: Formerly Concealed by the Deceased Dutchess of Lenox* (London, 1656), A3r. In recent years, printed books of secrets and recipe collections have received significant attention from literary scholars and historians alike. See, for example, Eamon, *Science and the Secrets of Nature*; Allison Kavey, *Books of Secrets: Natural Philosophy in England, 1550–1600* (Urbana: University of Illinois Press, 2007); Hunter, "Women and Domestic Medicine"; and Spiller, "Introduction," ix–li.

3. For surveys of medical print, see Mary Fissell, "Popular Medical Writing," in *Cheap Print in Britain and Ireland to 1660*, vol. 1 of *The Oxford History of Popular Print Culture*, ed. Joad Raymond (Oxford: Oxford University Press, 2011), 417–30; Fissell, "Marketplace of Print"; Paul Slack, "Mirrors of Health and Treasures of Poor Men: The Uses of the Vernacular Medical Literature of Tudor England," in *Health, Medicine and Mortality in the Sixteenth Century*, ed. Charles Webster (Cambridge: Cambridge University Press, 1979), 237–73; and Elizabeth Lane Furdell, *Publish-*

ing and Medicine in Early Modern England (Rochester, NY: University of Rochester Press, 2002).

4. From its debut in 1653 to the printing of the final edition in 1708, at least twenty-one editions of *A Choice Manual* were issued in a variety of formats, priced to suit all budgets. Nineteen editions (the latest edition printed in 1687) are recorded in Wing (K 310–17). John Ferguson, in his *Bibliographical Notes on Histories of Inventions and Books of Secrets*, 6th suppl., 30, mentions a twenty-first edition printed in 1708, a copy of which is held in the Wellcome Library. Similarly, the *STC* lists twenty entries from 1655 to 1698, with three additional printings of the "eleventh" edition in the eighteenth century (1710, 1713, and 1726).

5. All the printed recipe collections cited in this chapter were printed in London unless specified otherwise. All dates refer to publication of the first edition.

6. Wellcome, MS Western 3082, fol. 27r.

7. Fissell, "Marketplace of Print." On the increased commercializing and commodifying of medicine in early modern England, see, for example, Harold J. Cook's classic study, *The Decline of the Old Medical Regime in Stuart London* (Ithaca, NY: Cornell University Press, 1986). More recent studies include Patrick Wallis, ed., "Changes in Medical Care," *Journal of Social History*, special issue, 49, no. 2 (2016); Patrick Wallis, "Consumption, Retailing, and Medicine in Early-Modern London," *Economic History Review* 16 (2008): 26–53; and Ian Mortimer, *The Dying and the Doctors: The Medical Revolution in Seventeenth-Century England* (Woodbridge, UK: Royal Historical Society/Boydell Press, 2009). For a broader European view, see, for example, Harold J. Cook, *Matters of Exchange: Commerce, Medicine, and Science in the Dutch Golden Age* (New Haven, CT: Yale University Press, 2007).

8. Fissell, "Popular Medical Writing"; Fissell, "Marketplace of Print"; Slack, "Mirrors of Health."

9. Fissell, "Marketplace of Print," 116.

10. Leong, "Medical Recipe Collections," 38–41. For a recent overview of early modern English printed cookbooks, see Wall, *Recipes for Thought*, chap. 2.

11. As is described below, Narcissus Luttrell bought a copy of *The Accomplish'd Ladies Delight* for one shilling sixpence in 1675: Bodleian, Douce P 412. Using information from the Buxton papers in the Cambridge University Library, David McKitterick writes that the Buxtons bought a copy of *Ladies Cabinet Opened* in 1639 (the year the book was issued) for sixpence: McKitterick, "'Ovid with Littleton': The Cost of English Books in the Early Seventeenth Century," *Transactions of the Cambridge Bibliographical Society* 11, no. 2 (1997): 211.

12. According to the Buxton papers, an almanac cost twopence and play texts cost between sixpence and eight pence: McKitterick, "Ovid with a Littleton," 216, 223.

13. Printed recipe collections seem to have been sold at about four pence on the secondhand market. Nicholas Crouch, a fellow of Balliol College during the

late seventeenth century, bought a number of secondhand medical pamphlets or short treatises for between one pence and sixpence: Oxford, Balliol College 910.i.7.

14. Ferguson describes an 1815 copy of the work in his *Bibliographical Notes*. While he admits this differs somewhat from Lupton's original sixteenth-century work, the first ten books of the 1815 copy were nevertheless taken from the original. According to Ferguson, the work was also reprinted in 1826 by Griffin and Co., publishers of modern scientific works in Glasgow: Ferguson, part IV, 14–17.

15. Philibert Guybert's *The Charitable Physitian and the Charitable Apothecary* was first published in Lyon in 1623 as *Le médecin charitable enseignant la manière de faire et de préparer en la maison avec facilité et peu de frais les remèdes propres à toutes maladies, selon l'advis du médecin ordinaire.* The work went through several editions: see Laurence Brockliss and Colin Jones, *The Medical World of Early Modern France* (Oxford: Oxford University Press, 1997), 280; and Andrew Wear, "Popularized Ideas of Health and Illness in Seventeenth-Century France," *Seventeenth-Century French Studies* 8, no. 1 (1986): 229–42.

16. Fissell, "Marketplace of Print"; Webster, *Great Instauration,* chap. 4. For the impact of the Civil War on English printing, see Joad Raymond, *Pamphlets and Pamphleteering,* in *Early Modern Britain* (Cambridge: Cambridge University Press, 2006), esp. chap. 5; and Donald F. McKenzie, "The London Book Trade in 1644," in *Making Meaning: "Printers of the Mind" and Other Essays,* ed. Peter D. McDonald and Michael Felix Suarez (Amherst: University of Massachusetts Press, 2002), 126–43.

17. Leong, "Medical Recipe Collections," 43–44, 52; for a list of these titles, see appendix B.

18. A number of scholars have drawn attention to this; see, for example, Hunter, "Women and Domestic Medicine"; and Spiller, "Introduction."

19. William Lovell, *The Dukes Desk Newly Broken Up* (1660), title page.

20. For biographical information on Woolley, see John Considine, "Wolley, Hannah (*b.* 1622?, *d.* in or after 1674)," in *Oxford Dictionary of National Biography*; Elaine Hobby, *Virtue of Necessity: English Women's Writing, 1646–88* (London: Virago, 1988), chap. 7; and Elaine Hobby, "A Woman's Best Setting Out Is Silence: The Writings of Hannah Wolley," in *Culture and Society in the Stuart Restoration: Literature, Drama, History,* ed. Gerald Maclean (Cambridge: Cambridge University Press, 1995), 179–200.

21. A. T., *A Rich Store-house or Treasury for the Diseased* (1596), sig. A2v.

22. On the intersections of politics, science, and print, see Webster, *Great Instauration*; and Johns, *Nature of the Book,* chap. 4. For Culpeper, see also Mary Fissell, *Vernacular Bodies: The Politics of Reproduction in Early Modern England* (Oxford: Oxford University Press, 2004), chap. 5. For Culpeper's publishing, see F. N. L. Poynter, "Nicholas Culpeper and His Books," *Journal of the History of Medicine* 17 (1962): 152–67.

23. Fissell, "Marketplace of Print," 119–20.

24. Ralph Williams, *Physical Rarities: Containing the Most Choice Receipts of Physick and Chyrurgery* (1651), sig. A2r–v.

25. Owen Wood, *An Alphabetical Book of Physicall Secrets* (1639), 221. In the copy on *Early English Books Online*, this appears as page 217 because of a pagination mistake by the printer.

26. Ibid., sig. A3r.

27. Alexander Read or Reid (c. 1570–1641) was a surgeon and the author of several surgical treatises including *A Manuall of the Anatomy of the Body of Man* (1634) and *A Treatise of the First Part of Chirurgerie* (1638). A volume of recipes titled *Most Excellent and Approved Medicines*, said to have been compiled by Read, was published in 1652. Foster lists an Owen Wood from Angelsea who received a BA from Jesus College, Oxford, in 1580 and an MA in 1584. Wood went on to become dean of Armagh in 1590 and chaplain in ordinary to James I in 1606: Foster, *Alumni Oxonienses*, 1891.

28. Stafford's appended tract "The Physitians Help" was first published as *Helps for Suddain Accidents*, by Steven Bradwell (1633). Bradwell (1590/91–1646) came from a family of medical practitioners and was the grandson of the surgeon and author John Banister (1533–1599). He also wrote two treatises on the plague: see N. Gevitz, "'Helps for Suddain Accidents': Stephen Bradwell and the Origin of the First Aid Guide," *Bulletin for the History of Medicine* 67, no. 1 (1993): 51–73. For additional biographical information on John Banister, see Andrew Griffin, "Banister, John (1532/3–1599?)," in *Oxford Dictionary of National Biography*.

29. *STC* 25955 and *Wing* W 3404A–D. I have also identified two editions that are not listed in *Wing*, printed by John Stafford in 1658 and 1662. The former is in the British Library, and the latter is in the Wellcome Library.

30. See, for example, Robert Darnton, "What Is the History of Books?," in *The Kiss of Lamourette: Reflections of Cultural History*, ed. Robert Darnton (London: W. W. Norton, 1991), 107–36; Robert Darnton, "'What Is the History of Books?' Revisited," *Modern Intellectual History* 4 (2007): 495–508; Donald F. McKenzie, *Bibliography and the Sociology of Texts* (London: British Library, 1986); McDonald, McDonald, and Suarez, *Making Meaning*; Thomas Adams and Nicolas Barker, "A New Model for the Study of the Book," in *A Potencie of Life: Books in Society*, ed. Nicolas Barker (London: British Library, 1993), 5–43; and Adrian Johns, *The Nature of the Book: Print and Knowledge in the Making* (Chicago: University of Chicago Press, 1998).

31. Johns, *Nature of the Book*, chap. 1.

32. Other scholars have drawn our attention to this trend; see Hunter, "Women and Domestic Medicine"; and Spiller, "Introduction." There are, of course, also other instances of book producers' assigning a female author to a recipe book. *I secreti della signora Isabella Cortese* (1561) is a case in point. Historians and literary scholars alike have argued that "Isabella Cortese" likely is fictional, and the work demonstrates, at least as book producers saw things, the thirst for such books

compiled by women. For a detailed discussion, see Ray, *Daughters of Alchemy*, chap. 2.

33. Patricia Crawford, "Women's Published Writings, 1600–1700," in *Women in English Society, 1500–1800*, ed. Mary Prior (1985; repr., London: Methuen, 1996), 212–13.

34. For a biography of Grey, see Spiller, "Introduction," xxxi–xxxii.

35. For a biography of Howard and contemporary references of *Natura Exenterata* as her work, see ibid., xxv.

36. A. M., *Queen Elizabeths Closset of Physical Secrets* (1656), sig. A2r.

37. *Choice and Profitable Secrets both Physical and Chirurgical: Formerly Concealed by the Deceased Dutchess of Lenox* (London, 1656), sig. A3r–v.

38. Spiller, "Introduction," xxxii.

39. Madeline Bassnett, "Restoring the Royal Household: Royalist Politics and the Commonwealth Recipe Book," *Early English Studies* 2 (2009): 1–32; Knoppers, "Opening the Queen's Closet"; Jayne Archer, "The Queen's Arcanum: Authority and Authorship in the *Queens Closet Opened* (1655)," *Renaissance Journal* 1, no. 6 (2002): 14–26; Edith Snook, "'Soveraigne Receipts' and the Politics of Beauty in *The Queens Closet Opened*," *Early Modern Literary Studies*, special issue, 15 (2007): 1–19.

40. Bassnett, "Restoring the Royal Household."

41. Salvator Winter and Francisco Dickinson, *A Pretious Treasury, or A New Dispensatory* (London, 1649), sig. A2r.

42. *Natura Exenterata*, "Letter to the Reader," sig. A3r–v.

43. For an overview of vernacular medical literature and various kinds of empiricism, see Rankin, *Panaceia's Daughters*, chap. 1.

44. See Cook, *Decline of the Old Medical Regime*; Webster, *Great Instauration*, chap. 4; and Jenner and Wallis, *Medicine and the Market*.

45. See, for example, Cook, *Decline of the Old Medical Regime*; and Webster, *Great Instauration*, chap. 4. For a recent overview, see Lauren Kassell, "Secrets Revealed: Alchemical Books in Early Modern England," *History of Science* 49, no. 1 (2011): 74.

46. Margaret Ezell, "Cooking the Books, or The Three Faces of Hannah Woolley," in *Reading and Writing Recipe Books*, ed. Michelle DiMeo and Sara Pennell (Manchester: Manchester University Press, 2013), 167, 165. See also Hobby, *Virtue of Necessity*, chap. 7; and Hobby, "Woman's Best Setting Out Is Silence," 179–200.

47. This information stems from the medical part of the text. The often accompanying volume *Delight for Ladies* was not included in the count.

48. This translation of Oswald Gabelkover's *Artzneybuch* was first printed in Tübingen in 1594. The book was very popular (at least nine other editions still exist), and new editions appeared sporadically until the late seventeenth century. The work was translated into Dutch by Carel Batten in 1598 as *Medecyn-Boeck*

and published in Dordrecht. The same publishers issued the English translation in 1599.

49. This list is included in the second edition of *The Queens Closet Opened* (1658), sigs. B1r–2v. The *Natura Exenterata* also includes such a list. It may be that the book producers of *The Queens Closet Opened* followed suit: sig. A4r.

50. Rankin, *Panaceia's Daughters*, 86–87.

51. This practice can also be seen in other textual genres. For example, Deborah Harkness has pointed out that Hugh Plat, while carefully including source information in his manuscript writings, omitted the names of his informants for many experiments in his printed book *The Jewell House*: Harkness, *Jewel House*, 240–41.

52. Ezell, *Social Authorship*, 38–39.

53. These recipes apparently were requested by and sent to Dr. William Avery in New England: Michael Hunter, "The Reluctant Philanthropist: Robert Boyle and the 'Communication of Secrets and Receits in Physick,'" in *Religio Medici: Medicine and Religion in Seventeenth-Century England*, ed. Ole Peter Grell and Andrew Cunningham (Aldershot, UK: Ashgate, 1996), 247–72.

54. Ibid., 262–63.

55. Boyle, *Some Receipts of Medicines* (1688), sig. A4r–v. Michelle DiMeo also discusses this system in "'Communicating Medical Recipes': Robert Boyle's Genre and Rhetorical Strategies for Print," in *The Palgrave Handbook to Early Modern Literature and Science*, ed. Howard Marchitello and Evelyn Tribble (Basingstoke, UK: Palgrave Macmillan, 2017), 209–28.

56. Boyle, *Some Receipts of Medicines*, sig. A4r–v; DiMeo, "'Communicating Medical Recipes,'" 220.

57. Scholars who study scribal publication have pointed out that many private manuscripts, though they circulated, did so within restricted circles. Margaret Ezell terms this type of manuscript circulation "social publication": Ezell, *Social Authorship*, chap. 1. See also Marotti, *Manuscript*, 30–47; and Love, *Scribal Publication*, esp. chap. 5. On manuscript culture among early modern British naturalists, see Yale, *Social Knowledge*, and Yale, "Marginalia." In her study of John Evelyn's reading John Aubrey's *Naturall Historie*, Yale notes that Evelyn also subtracted information about the source of particular natural knowledge: "Marginalia," 199.

58. The history of reading is a rich and well-developed field. Pioneering studies include Roger Chartier, *The Order of Books: Readers, Authors, and Libraries in Europe between the 14th and 18th Centuries* (Stanford, CA: Stanford University Press, 1994); and Robert Darnton, "First Steps toward a History of Reading," *Australian Journal of French Studies* 23, no. 1 (1986): 5–30. For collections of essays, see Kevin Sharpe and Steven Zwicker, eds., *Reading, Society and Politics in Early Modern England* (Cambridge: Cambridge University Press: 2003); James Raven, Helen Small, and Naomi Tadmor, eds., *The Practice and Representation of Reading in England* (Cambridge: Cambridge University Press, 1996); Jennifer Andersen and Elizabeth Sauer, eds., *Books and Readers in Early Modern England: Material*

Studies (Philadelphia: University of Pennsylvania Press, 2002); and, more recently, Jennifer Richards and Fred Schurink, "The Textuality and Materiality of Reading in Early Modern England," *Huntington Library Quarterly*, special issue, 73, no. 3 (2010): 345–61.

59. Lisa Jardine and Anthony Grafton, "'Studied for Action': How Gabriel Harvey Read His Livy," *Past and Present* 129, no. 1 (1990): 30–78; Kevin M. Sharpe, *Reading Revolutions: The Politics of Reading in Early Modern England* (New Haven, CT: Yale University Press, 2000); William Howard Sherman, *John Dee: The Politics of Reading and Writing in the English Renaissance* (Amherst: University of Massachusetts Press, 1995).

60. Lisa Jardine and William Sherman, "Pragmatic Readers: Knowledge Transactions and Scholarly Services in Late Elizabethan England," in *Religion, Culture and Society in Early Modern Britain*, ed. Anthony Fletcher and Peter Roberts (Cambridge: Cambridge University Press, 1994), 102–24.

61. For an account of "humble" readers and their reading practices, see Margaret Spufford, "First Steps in Literacy: The Reading and Writing Experiences of the Humblest Seventeenth-Century Autobiographers," *Social History* 4 (1979): 407–35; and Spufford, *Small Books and Pleasant Histories: Popular Fiction and Its Readership in Seventeenth-Century England* (Cambridge: Cambridge University Press, 1981).

62. Brayman Hackel, *Reading Material in Early Modern England: Print, Gender, and Literacy* (Cambridge: Cambridge University Press, 2005); Lori Humphrey Newcomb, *Reading Popular Romance in Early Modern England* (New York: Columbia University Press, 2002).

63. Harkness, *Jewel House*; Kassell, "Secrets Revealed"; Renée Raphael, *Reading Galileo: Scribal Technologies and the Two New Sciences* (Baltimore: Johns Hopkins University Press, 2017); Yale, *Sociable Knowledge*.

64. For a discussion of how to read recipes, see Ezell, "Domestic Papers"; Theophano, *Eat My Words*, chap. 5; Wendy Wall, "Literacy and the Domestic Arts," *Huntington Library Quarterly* 73, no. 3 (2010): 383–412; and Wall, *Recipes for Thought*, chap. 1.

65. Grafton and Jardine, "Studied for Action"; Sherman, *Used Books*.

66. Literature on readers' annotations is wide-ranging. The seminal article in the field remains Jardine and Grafton, "Studied for Action." For an overview of early modern English readers' annotations, see Sherman, *Used Books*; and William H. Sherman, "What Did Renaissance Readers Write in Their Books?," in *Books and Readers in Early Modern England: Material Studies*, ed. Jennifer Andersen and Elizabeth Sauer (Philadelphia: University of Pennsylvania Press, 2002), 119–37.

67. Bodleian, Vet. A3 f.1872. Milne, whom I have been unable to identify, seems to have made several attempts at writing his name. Unfinished inscriptions can be seen on the recto and verso of the front flyleaf and on sigs. A1r, A2v, A5v, and 194. Luttrell's annotated copy of *The Accomplish'd Ladies Delight* is Bodleian, Douce

P 412. The practice of writing one's name multiple times in a volume is described in Hackel, *Reading Material*, 160.

68. The book was published in 1666, and Milne annotated his copy sometime during 1743.

69. Bodleian, Douce R 66 and Douce C 25 (1).

70. Folger, 138-469q, 170 and back flyleaf verso.

71. Bodleian, Tanner 880, title page and front flyleaf.

72. Folger, *STC* 296 copy 2, *STC* 305 copy 2, and *STC* 301 copy 2, part 3, fols. 36v, 57v, 60v, 62v.

73. Bodleian, Locke 7.411.

74. Folger, 151-446q, front flyleaf recto and back flyleaf recto.

75. Huntington, HN Rare Book 389398.

76. Ibid., 115, 109.

77. Folger, 11798 copy 3; examples can be seen on 129v–130v and 224v.

78. Bodleian, Vet. A2. f.1, 61, 161.

79. Bodleian, Vet. A1 e.102, 27, 248.

80. Bodleian, 20 d.6, 85.

81. Bodleian, Tanner 212; Jones, "Formula and Formulation."

82. Bodleian, 20 d.6, 85.

83. Bodleian, Vet. A3 f.1788, 205, 5 (of *The Compleate Cook*).

84. This is a copy of the first (1653) edition of Elizabeth Grey's *A Choice Manual*. The volume is otherwise unmarked: Bodleian, 8° C 571 Linc, 102–4. According to a contemporary guide to drugs, Lapis contrayerva "is a made Stone of contrayerva-Roots, Pearls, Coral, Amber, Crabs-eyes, and several other Ingredients, which after is gilded over, smelleth strong of the Root": John Jacob Berlu, *The Treasure of Drugs Unlock'd* (London: John Harris and Tho. Howkins, 1690), 64. The *OED* defines contrayerva as "a name given, in general use, to the root-stock and scaly rhizome of species of *Dorstenia* (*D. contrayerva* and *D. braziliensis*), family Urticaceae native to tropical America, used as a stimulant and tonic, and formerly as an antidote to snake-bites. In Jamaica, the name is given to a species of Birthwort (*Aristolochia odoratissima*), still held in repute as an alexipharmic."

85. Bodleian, Tanner 212 (2), fol. 33v.

86. For example, the reader of Bodleian RR z. 179 (a copy of Jean Goeurot's *The Regiment of Life* [1560]) wrote in a recipe for the "coldness of the liver" on the title page. Bodleian G. Pamph. 2156 (1), a copy of *A Hospitall for the Diseased*, has several additional recipes (now faded) written in on the back page. Examples of readers' inserting recipes addressing the same ailment into blank spaces near the printed recipes can be seen in Bodleian Vet. A1 e.4, a 1596 edition of Peter Leven's *A Right and Profitable Booke for All Diseases, Called the Pathway to Health*, and BL, 1038.i.35 (7)/1038.k.47 (3), a copy of *A Rich Store-house* with the ownership note of Matthew Wattes, dated 1610. The volume is heavily annotated by Wattes, who took a number of recipes from "Tho Eaton's book," which I assume is a contemporary manuscript.

87. Huntington, Rare Book 53930. The reader bound sixteen extra pages into the front and back of the volume, and about fifty additional recipes were written in. There is also a section on the medicinal virtues of herbs at the back of the volume.

88. Huntington, Rare Book 124144 and HM MS 59539 (annotations from the volume).

89. The literature on learned note taking is vast. For an excellent overview, see Blair, *Too Much to Know.*

90. For an in-depth discussion of Margaret Boscawen's medical reading habits, see Leong, "Herbals She Peruseth."

91. Devon Record Office 1262 M/FC/6. The manuscript is unfoliated; this section occurs in the middle of the manuscript.

92. For a detailed study of Elizabeth Freke's medical and reading practices, see Leong, "Herbals She Peruseth"; and Leong, "Making Medicines." For a general study of early modern women reading herbal texts, see Laroche, *Medical Authority.*

93. Wellcome, MS Western 2630.

94. NLM, MS b 261 and Folger MS v.a. 452. For a detailed discussion on Greenway's reading practices, see Leong, "Medical Recipe Collections," chap. 4. I detail the story of Rivière's *Praxis Medica* and *Observationes* in England in Leong, "Printing Vernacular Medicine in Early Modern England: The Case of Lazare Rivière's *The Practice of Physick*," in *Civic Medicine: Physician, Polity and Pen in Early Modern Europe*, ed. J. Andrew Mendelsohn, Annemarie Kinzelbach, and Ruth Schilling (London: Routledge, 2018). Other manuscript recipe books with extracts or references to Rivière's *Observationes*, which in England circulated with the translation of his *Praxis Medica*, include CUL, MS Dd 2 45 (notebook of Dru Burton with notes and translations from Lazare Rivière and others); Glasgow, MS Hunter 485 and MS Hunter 487.

95. On commonplacing, see Ann Moss, *Printed Commonplace-Books and the Structuring of Renaissance Thought* (Oxford: Clarendon Press, 1996); Stolberg, "Medizinische Loci Communes"; and Blair, *Too Much to Know.*

96. Gervase Markham, *The English Hus-wife* (London: Roger Jackson, 1615), sig. Q1v. Though with its own title page, this edition was issued as part of *Country Contentments* (London, 1615).

97. W. M., *The Queens Closet Opened* (London: Nathaniel Brooks, 1655), sigs. A3v–A4r.

98. Grey, *Choice Manual*, 102. See note 84 above for more information on lapis contrayerva or yerba. For in-depth discussions of this book, see Hunter, "Women and Domestic Medicine"; and Spiller, "Introduction."

99. Greengrass, Leslie, and Hannon, *Hartlib Papers*, accessed 19 August, 2017, https://www.hrionline.ac.uk/hartlib, 29/4/19A.

100. BL, MS Additional 45196, fol. 76v.

101. Wellcome, MS Western 4338, fols. 138r, 141r; BL, MS Additional 56248, fol. 104v; Folger, v.a. 215, 111. The recipe can also be found, in various guises often

passed on by an intermediary, in numerous collections across the recipe archive. See, for example, RCP, MS 497m fol. 40v; Wellcome, MS Western 2990, fol. 109v; Folger, MS v.a. 361, fol. 169v; and BL, MS Additional 30164, fol. 37v.

102. BL, MS Stowe 1077, fol. 11v.

103. *The Queens Closet Opened* (London, 1655), 150–52.

104. Ibid., 153.

105. Wellcome, MS Western 160, 121; MS Western 3082, fol. 165r; MS Western 3712, fols. 76v–77r; MS 4338, fol. 207r–v. Other manuscript collections with a version of this recipe include Wellcome, MS Western 774, 31, and MS Western 3712, fols. 76r–77v; Palmer, *Recipe Book*, 85; and Folger MS v.a. 430, 59.

106. For Packe, see Folger, MS v.a. 215, 109; and *A Choice Manual* (London, R. Norton, 1653), 100–101.

107. Wellcome, MS Western 7301, 71; MS Western 4338, fols. 138r, 141r.

108. NYAM, "[a] collection of choice receipts," 130, 134. For more on the St. John family, see Frank T. Smallwood, "The Will of Dame Johanna St. John," *Notes and Queries* 16, no. 9 (1969): 346.

109. NLM, MS f 98, fol. 1r–v.

110. Ibid., fol. 2r–v.

111. Ibid., fol. 3v.

112. For a discussion on the common tripartite structure of printed recipe books, see Hunter, "Women and Domestic Medicine."

113. As examples, *A Choice Manual* (1653); *The Queens Closet Opened* (1655); Salvator Winter and Francisco Dickinson, *A Pretious Treasury, or A New Dispensatory* (1649); and Brugis, *Marrow of Physick* (1640).

114. Elizabeth Yale has written extensively on how naturalists viewed print and manuscript texts in different lights. Whereas handwritten papers on natural history could be continually expanded to accommodate new knowledge, print works "seemed to impose a finality": *Sociable Knowledge*, 9.

115. The classic studies for this are Love, *Culture and Commerce of Texts*; and Marotti, *Manuscript, Print, and the English Renaissance Lyric.* For a recent study of this issue, focused on the history of science, see Yale, *Sociable Knowledge.*

116. See also Herbert, *Female Alliances*, chap. 3; and Wall, *Recipes for Thought*, chap. 1.

117. Harkness, *Jewel House*; Kassell, *Medicine and Magic*; Yale, *Sociable Knowledge.*

Conclusion

1. Carne, *Some St. John Family Papers*, 47.

2. On the History of Trades program, see Michael Hunter, *Science and Society in Restoration England* (Cambridge: Cambridge University Press, 1981), chap. 4;

Walter E. Houghton, "The History of Trades: Its Relation to Seventeenth-Century Thought; as Seen in Bacon, Petty, Evelyn, and Boyle," *Journal of the History of Ideas* 2, no. 1 (1941): 33–60; and Kathleen H. Ochs, "The Royal Society of London's History of Trades Programme: An Early Episode in Applied Science," *Notes and Records of the Royal Society of London* 39, no. 2 (1985): 129–58. On the French Academy of Sciences' testing and trying of mineral spa waters, see Michael Bycroft, "Iatrochemistry and the Evaluation of Mineral Waters in France, 1600–1750," *Bulletin of the History of Medicine* 91, no. 2 (2017): 303–30.

3. On wonder and curiosity, see, for example, Lorraine Daston and Katharine Park, *Wonders and the Order of Nature, 1150–1750* (Cambridge, MA: Zone Books, 2001); Robert John Weston Evans and Alexander Marr, *Curiosity and Wonder from the Renaissance to the Enlightenment* (Aldershot, UK: Ashgate, 2006).

Bibliography

Manuscripts

I have viewed all the manuscripts included in this list, and for the most part the descriptions given are my own. Ownership notes generally refer to the formulaic phrase "X X: her/his book, date." The dates in parentheses refer to those stated in the ownership note. "Associated with" and "connected with" refer to collections that had sections "taken out" of the notebook of a particular person. Where "recipe collection of" is used, the compiler of the collection was identified through internal evidence. Although it has not been possible to describe all the types of manuscripts and printed texts that were bound or copied with particular recipe collections, I have used "medical miscellany" and "medical notebook" to describe manuscripts where the recipe collection was part of a volume of different texts. "Commonplace book" here refers to volumes of Latin *sententiae* that were organized under strict headings. Where the compiler of a manuscript collection copied long extracts from a particular printed text, I have provided the title of the text. However, owing to the number of books cited, this information has not been included, for example, in the entries on Elizabeth Freke and "cousin Greenway."

Manuscript Collections in the United Kingdom

Cambridge, Cambridge University Library
    MS Additional 3071 [recipe collection with ownership notes of Philipe Humpry and Thomas Spencer]
    MS Dd 2 28 [anonymous recipe collection]
    MS Dd 2 34 ["Sondry remedies collected out of the Dr Dan Sennertus his workes, and put into English by D.B. for his owne p[ar]ticular use" (D.B. is Dru Burton)]

MS Dd 2 41 [recipe collection of Dru Burton with extracts taken from Lazarus Riverius and Amatus Lusitanus]

MS Dd 2 45 [medical notebook of Dru Burton]

MS Dd 5 1 [anonymous recipe collection]

MS Dd 5 18 [recipe collection with ownership note of John Gardiner (1652)]

MS Dd 5 66 [recipe collection connected with the Langley family]

MS Dd 5 68 [recipe collection connected with the Corbett family]

MS Dd 6 9 [anonymous recipe collection]

MS Dd 6 79 [anonymous recipe collection]

MS Dd 8 36 ["book of phisicall Receipts" with ownership note of Henry Harcourt (1649)]

MS Dd 12 38 [recipe collection with ownership note of Alexander Ross]

MS Dd 12 60 [recipe collection with ownership note of Ed. Webb (1657)]

MS Ll 5 8 ["A very good collection of approved recipes of chymycal operations collected by Augustus Kuffeler and Charles Ferrers phylchimist. Anno Domini 1666"]

MS Oo 7 46 [anonymous recipe collection with sections taken out of *The Queens Closet Opened*]

Devon Record Office

1262 M/FC/6 [recipe book of Margaret Boscawen and Bridget Fortescue]

1262 M/FC/7 [plant notebook of Margaret Boscawen and Bridget Fortescue]

Glasgow, University of Glasgow Library

MS Ferguson 43 ["A collection of sundrie approved receipts in phisike & chirurgerie"]

MS Ferguson 61 [recipe collection bearing the ownership note of Mary Harrison (1692)]

MS Ferguson 150 [anonymous recipe collection with recipes in Italian and French]

MS Gen 831 [anonymous recipe collection]

MS Hunter 64 [anonymous recipe collection]

MS Hunter 93 [John de Feckenham, "Booke of Soveraigne Medicines"]

MS Hunter 95 [medical notebook with recipes]

MS Hunter 169 [Henry Fowler, "Fasciculus"]

MS Hunter 243 [medical notebook including extracts from *Pharmacopoeia Londinensis*]

MS Hunter 485 [anonymous recipe collection]

MS Hunter 487 [anonymous recipe collection]

London, British Library

MS Additional 27466 [recipe collection bearing the ownership note of Mary Doggett (1682)]

MS Additional 28320 [anonymous recipe collection]

MS Additional 28327 [anonymous recipe collection with ownership notes of Margaret Yelverton, Katherine Stephens, and Julian Skimington]

MS Additional 28643 [recipe collection including extracts taken out of *The Secretes of the Reverend Maister Alexis of Piemont*]

MS Additional 28956 [anonymous recipe collection]

MS Additional 29306 [recipe collection connected with John Barrington]

MS Additional 30164 [anonymous recipe collection]

MS Additional 30502 [recipe collection connected with the Dacre family]

MS Additional 34210 [anonymous recipe collection]

MS Additional 34212 [Thomas Lodge, "The Poor Mans Talent"]

MS Additional 34722 [recipe collection associated with Lady Anne Lovelace and bearing the ownership note of Cisilia Haynes (1659)]

MS Additional 35342 [notebook used as a medical and culinary recipe collection and a commonplace book]

MS Additional 36308 [recipe collection belonging to multiple owners including Dorothy Washborn]

MS Additional 37719 [commonplace book with medical recipes bearing the ownership note of John Gibson (1656)]

MS Additional 42115 [anonymous recipe collection]

MS Additional 45193 [commonplace book with genealogical notes on the Stoughton family of Stoughton Place with annotations in multiple hands]

MS Additional 45195 [notebook of Richard Glyd]

MS Additional 45196 [recipe collection bearing the ownership note of Anne Glyd (1656)]

MS Additional 45197 [recipe book bearing the ownership note of Anne Brockman (1638)]

MS Additional 45198 [loose recipes associated with the Brockman family]

MS Additional 45199 [recipe book bearing the ownership note of Elizabeth Brockman (1674)]

MS Additional 45120 [notebook associated with James Brockman]

MS Additional 45718 [remembrances and recipe collection of Elizabeth Freke]

MS Additional 48193 [anonymous recipe collection with extracts from *The Queens Closet Opened*]

MS Additional 52585 [commonplace book with medical recipes associated with Richard Waferer]

MS Additional 56248 [recipe collection bearing the ownership note of Mary Dacre (1666)]

MS Additional 56279 [commonplace book associated with Sir Roger Aston (d. 1612), containing medical recipes]

MS Additional 69409 [anonymous recipe collection]

MS Additional 69970 [anonymous recipe collection dated 1689]

MS Additional 70006 [Harley family correspondence and papers]

MS Additional 70007 [Harley family correspondence and papers]

MS Additional 70113 [Portland papers]

MS Additional 72446 [recipe collection bearing the ownership note of
    Elizabeth Trumbell]
MS Additional 72619 [recipe collection associated with the Trumbell family]
MS Additional 78337 [recipe collection of John and Mary Evelyn]
MS Egerton 2197 [recipe collection bearing the ownership note of and
    compiled by Elizabeth Digby (1650)]
MS Egerton 2214 [anonymous recipe collection]
MS Egerton 2415 [recipe collection bearing the ownership note of Mary
    Birkhead (1681)]
MS Egerton 2203 [anonymous recipe collection]
MS Harley 2389 [anonymous recipe collection]
MS Harley 6490 [anonymous recipe collection copied into a commonplace
    book]
MS Harley 6816 [miscellaneous notebook including medical recipes]
MS Sloane 556 [recipe collection associated with "La Maques Dorsetts"]
MS Sloane 1289 [recipe collection associated with "Madam Porter," "Mad.
    Jones," and the "Lady Berklys family" (1681)]
MS Sloane 1367 [medical recipe collection titled "My Lady Rennelaghs
    Choise Receipts, as also some of Capt. Willis who valued them above
    gold"]
MS Sloane 1468 [notebook with miscellaneous medical texts including
    "Margaret Crux her book of Receipts"]
MS Sloane 2266 [recipe collection connected with the Grosvenor family]
MS Sloane 2485/6 [recipe collection connected with Lettis Corbett and
    Margaret Baker (1650)]
MS Sloane 2488 [recipe collection bearing the ownership note of Eliza-
    beth Beere]
MS Sloane 3235 [recipe collection bearing the ownership note of Mary
    Grosvenor (1649)]
MS Sloane 3842 [recipe collection with the ownership note of Hester Gul-
    lyford and a presentation note from "Poore Colly" to Elizabeth Butler
    (1679)]
MS Stowe 1077 [recipe collection of Sir Peter Temple and Elinor Temple]
MS Stowe 1078 [recipe collection of Sir Peter Temple bearing the owner-
    ship note of Elinor Temple]
MS Stowe 1079 [recipe collection of Sir Peter Temple]
London, National Archives
    State Papers, Domestic, SP 18/11
    State Papers, Domestic, SP 18/16
London, Royal College of Physicians Library
    MS 232 ["A Booke of Physicall Receipts Worth the Observing and Keep-
        ing for Mrs Alice Corffilde 1649"]

MS 447 [anonymous recipe collection]

MS 497 [anonymous recipe collection]

MS 498 [anonymous recipe collection]

MS 499 [anonymous recipe collection]

MS 500 [recipe collection bearing the ownership notes of Edmond Quarles and Susan Pytches]

MS 501 [anonymous recipe collection]

MS 502 [anonymous recipe collection]

MS 503 [anonymous recipe collection including physicians' prescriptions]

MS 504 [anonymous recipe collection perhaps associated with the Clark family]

MS 505 [anonymous recipe collection]

MS 506 [anonymous recipe collection, taken "out of my Aunt Bromfeilds book"]

MS 509 [recipe collection of the Acton family]

MS 513 [recipe collection bearing the ownership note of Johannes Hussey (1667)]

MS 534 [recipe collection bearing the ownership note of Lady Sedley (1686)]

MS 654 [recipe collection bearing the ownership note of Sarah Wigges (1616)]

MS 688 [recipe collection bearing the ownership note of Honore Henslow (1601)]

London, Royal College of Surgeons Library

MS 42.d.18 [recipe collection bearing the ownership note of Elizabeth Isham (1659)]

MS 153 K 3 [recipe collection bearing the ownership note of Sydney Humphryes]

MS Additional 112 [anonymous recipe collection]

London, Wellcome Library

MS Western 1 [recipe collection bearing the ownership note of Grace Acton (1621)]

MS Western 108 [recipe collection bearing the ownership note of Jane Baber]

MS Western 143 ["A Booke of receipts both for Phisicke and Chirurgery gathered from divers excellent practitioners and sundry Authors 1641" with the ownership note of Edward Petchey (1690)]

MS Western 144 [anonymous recipe collection]

MS Western 159 [medical commonplace book of Edward Bruer]

MS Western 160 [recipe collection bearing the names of Anne Brumwich, Rhoda Fairfax (1663), Ursula Fairfax, and Dorothy Cartwright]

MS Western 169 [recipe collection bearing the ownership note of Elizabeth Bulkeley (1627)]

MS Western 184a [recipe collection of Lady Frances Catchmay]

MS Western 212 [recipe collection bearing the ownership notes of Arthur and Alice Corbett]

MS Western 213 [recipe collection bearing the ownership notes of the Countess of Arundel and Mrs. Corlyon (1606)]

MS Western 311 [recipe collection bearing the ownership notes of John (1634), Joan (1669), and Joanna Gibson (1708)]

MS Western 363 [recipe collection of Sarah Hughes]

MS Western 373 [recipe collection bearing the ownership note of Jane Jackson (1642)]

MS Western 579 [recipes collected by George Noble (1629)]

MS Western 635 [anonymous recipe collection]

MS Western 676 [anonymous recipe collection]

MS Western 716 [recipe collection with an iatrochemical focus including recipes by William Ruthven]

MS Western 749 [collection of fourteen manuscripts once belonging to James Shrowl (1814–97), including several medical recipe collections]

MS Western 751 [recipe collection of Elizabeth Sleigh and Felicia Whitfeld]

MS Western 761–2 [recipe collection of Philip Stanhope, first Earl of Chesterfield]

MS Western 768 [anonymous recipe collection including extracts from A. T., *A Rich Store-house*]

MS Western 774 [recipe collection bearing the ownership note of Dorcas Gwynn (1636)]

MS Western 796 [recipe collection bearing the ownership note of Thomas Walker]

MS Western 809 [anonymous recipe collection]

MS Western 1026 [recipe collection with the ownership note of Lady Ayscough (1692)]

MS Western 1071 [recipe collection bearing the ownership notes of Ann Egerton and "Lady Barrett"]

MS Western 1098 [recipe collection with the ownership notes of Thomas and Mary Bathurst]

MS Western 1322 [anonymous recipe collection]

MS Western 1323 [anonymous recipe collection]

MS Western 1325 [anonymous recipe collection]

MS Western 1340 [recipe collection associated with the Boyle family]

MS Western 1364 [recipe collection associated with Joseph Brooker]

MS Western 1511 [recipe collection bearing the ownership note of Madam Carr and dated to 1681/82]

MS Western 1548 [recipe collection bearing the ownership note of Mrs. Mary Chantrell (1690)]

MS Western 1662 [recipe collection bearing the ownership notes of Mary
    Clerke and Barbara Dicke]
MS Western 1710 [recipe collection associated with William Coleman
    (1657)]
MS Western 2031 [recipe collection associated with Nathaniel Dalton
    (1668)]
MS Western 2287 [anonymous recipe collection with miscellaneous mate-
    rials including an astromedical text]
MS Western 2323 [recipe collection with the ownership note of Amy
    Eyton]
MS Western 2330 [anonymous recipe collection]
MS Western 2535 [recipe collection of Elizabeth Godfrey (1686)]
MS Western 2630 [Extracts from *A Choice Manual of Rare and Select Se-
    crets*, eighteenth century]
MS Western 2840 [recipe collection bearing the ownership note of Elizabeth
    Hirst (1684)]
MS Western 2844 [recipe collection of Martha Hodge]
MS Western 2954 [recipe collection bearing the ownership note of Sarah
    Hudson (1668)]
MS Western 2990 [recipe collection bearing the ownership note of Bridget
    Hyde (1676)]
MS Western 3009 [recipe collection bearing the ownership note of Eliza-
    beth Jacob (1654)]
MS Western 3082 [recipe collection with the ownership note of Elizabeth
    Philips (1694)]
MS Western 3107 [recipe collection associated with Edward and Kather-
    ine Kidder (1699)]
MS Western 3341 [recipe collection bearing the ownership note of
    W[illiam] Lowther]
MS Western 3500 [recipe collection associated with Mrs. Meade and others]
MS Western 3547 [recipe collection bearing the ownership note of Mrs.
    Mary Miller/Criche (1660)]
MS Western 3712 [recipe collection associated with Elizabeth Okeover
    Adderley and others]
MS Western 3768 [recipe collection bearing the ownership note of
    Bridgett Parker (1663)]
MS Western 3769 [recipe collection bearing the ownership note of Jane
    Parker (1651)]
MS Western 3834 [recipe collection bearing the ownership note of Kather-
    ine Perrott and Mary Perrott (1695)]
MS Western 4047 [anonymous recipe collection]

MS Western 4048 [anonymous recipe collection]

MS Western 4049 [anonymous recipe collection]

MS Western 4050 [anonymous recipe collection]

MS Western 4051 [anonymous recipe collection with household accounts
   (1695)]

MS Western 4054 [anonymous recipe collection]

MS Western 4338 [recipe collection bearing the ownership note of Johanna St. John (1680)]

MS Western 4683 [recipe collection with the ownership note of Frances
   Springatt (1686)]

MS Western 4887 [recipe collection of James Tyrrell (1642–1718), with sections copied from the Hartlib papers in Brereton Hall]

MS Western 5093 [commonplace book with medical recipes belonging to
   the Woodward family]

MS Western 7391 [recipe collection associated with the Okeover family]

MS Western 7113 [recipe collection of Anne Fanshawe (1651)]

MS Western 7849 [recipe collection bearing the ownership note of Theo-
   dosia Henshare]

MS Western 7851 [recipe collection with the ownership notes of Mary
   Dawes (1791), Penelope Humphries, and Elizabeth Browne (1697)]

MS Western 7997 [Heppington Receipts volume 1]

MS Western 7998 [Heppington Receipts volume 2]

MS Western 7999 [Heppington Receipts volume 3]

MS Western 8000 [recipe collection associated with the Godfrey/Faussett
   family]

MS Western 8001 [recipe collection associated with the Godfrey/Faussett
   family]

MS Western 8002 [recipe collection associated with the Godfrey/Faussett
   family]

MS Western 8680 [recipe collection associated with the Godfrey/Faussett
   family]

MS Western MS/MSL/2 [anonymous recipe collection]

Oxford, Bodleian Library

MS Ashmole 321 [collection of medical tracts, including medical recipes,
   bearing the ownership notes of Francis Park (1636), Th. Broghton, and
   William Lilly]

MS Ashmole 1463 [notebook of miscellaneous information including recipes]

MS Ashmole 1489 [collection of four books of medical recipes associated
   with Richard Saunders and bearing the ownership note of and "made by"
   Matthew Barrett (1626) and the ownership note of Elias Ashmole]

MS Aubrey 19 [recipe collection with the ownership note of R. Carrow]

MS Don. c. 24 [recipe collection with ownership note of Margaret Man]

MS Don. e. 11 [recipe collection of the Carey family]

MS Don. e. 6 [commonplace book with recipes with the ownership notes
    of Katherine and William Cartwright]
MS Eng. Hist. c. 250 [anonymous recipe collection]
MS Eng. Hist. e. 199 [recipe collection with the ownership note of Mary
    Widdrington]
MS Eng. Misc. d. 436 [recipe collection with the ownership notes of Anne
    and Elizabeth Bertie]
MS Eng. Misc. d. 437 [loose recipes associated with Anne and Elizabeth
    Bertie]
MS Lincoln College 127 [recipe collection with sections written for Dr. Herd]
MS Lincoln College lat. 122 [medical miscellany collected by John Herd]
MS Lister 5 [recipe collection within the papers of Martin Lister]
MS Rawl. c. 81 [recipe collection bearing the ownership note of Dorothy
    Hudson (1629)]
MS Rawl. d. 699 [recipe collection with the ownership notes of Gregory
    Pyford, Edward Stanhope, Thomas Tydley, and William Anderson]
MS Rawl. d. 947 [commonplace book and recipe collection with the own-
    ership notes of Jo Candyne and Thomas Keen]
MS Rawl. d. 1056 [recipe collection with the ownership note of Edward
    Bastard]
MS Rawl. d. 1074 [anonymous recipe collection]
MS Rawl. d. 1447 [anonymous commonplace book and recipe collection]
MS Sancroft 120 [recipe collection of William Sancroft (1617–93), archbishop
    of Canterbury, partly from Dr. William Sancroft the elder's papers]
MS Tanner 397 [recipe collection of Valentine Bourne (1610)]
MS Top Oxon. c. 757 [notebook of the Napier family, including medical
    recipes]
MS Top Oxon e. 380 [anonymous recipe collection]
Oxford, Exeter College Library
    MS 71 [anonymous recipe collection]
    MS 78 [recipe collection with the ownership note of Johnny Sheffeilde (1650)]
    MS 84 [anonymous recipe collection]
Northampton, Northampton County Record Office
    Westmorland (Apethorpe) Collection—Additional Items
    W (A) misc. vol. 32–33 [three volumes of medical papers connected with
        Lady Grace Mildmay and her daughter Mary Fane, Countess of West-
        morland]
    MS FH 4249 [anonymous collection of loose recipes]
Wiltshire, Wiltshire and Swindon History Centre
    MS 3430 [Papers of the St. John family of Lydiard Park]
Worthing, Worthing County Museum
    MS 3574 [recipe collection associated with the household of the Dukes of
        Norfolk]

Manuscript Collections in the United States

Bethesda, Maryland, National Library of Medicine
 MS b 131 [anonymous recipe collection]
 MS b 174 [anonymous recipe collection]
 MS b 256 [anonymous recipe collection]
 MS b 259 [anonymous recipe collection bound with a French grammar book]
 MS b 261 [recipe collection connected with "cousin Greenway"]
 MS c 349 [anonymous recipe collection]
 MS f 98 [anonymous recipe collection]
New York City, Manuscripts and Archives Division, New York Public Library,
 Astor, Lenox, and Tilden Foundations
 English household cookbook [recipe collection bearing the ownership
  note of Mrs. Anne Meyricke]
 Whitney Cookery, Collection MS 2 [recipe collection written in the hand
  of Lady Anne Percy]
 Whitney Cookery, Collection MS 4 [recipe collection bearing the owner-
  ship note of Lady Anne Morton (1693)]
 Whitney Cookery, Collection MS 5 [recipe collection bearing the owner-
  ship notes of Mary Davies and Lettice Cotton (1684)]
 Whitney Cookery, Collection MS 6 [anonymous recipe collection]
 Whitney Cookery, Collection MS 8 [anonymous recipe collection]
 Whitney Cookery, Collection MS 9 [recipe collection of the Bennett family]
 Whitney Cookery, Collection MS 10 [anonymous recipe collection]
 Whitney Cookery, Collection MS 11 [recipe collection bearing the owner-
  ship note of Hester Denbigh (1700)]
New York City, New York Academy of Medicine Library
The descriptions given are the actual classmarks.
 MS approved receipts in physicke
 MS collection of choice recipes
 MS A collection of medical remedies for colic and other—cover title Old
  Doctor 1690
 MS Owen Salesbury
 MS receipts for medical remedies
 MS receipts for medical remedies
San Marino, California, Henry E. Huntington Library
 HM MS 41536 [commonplace book with recipes by Sir Edward Dering]
 HM MS 59539 [annotations in Thomas Brugis, *Vade Mecum*]
 HM MS 60413 [recipe collection associated with the Danby family]
Washington, DC, Folger Shakespeare Library
 MS e.a. 4 [commonplace book with medical recipes]
 MS e.a. 5 [medical miscellany c. 1634 with some lectures by the members
  of the Worshipful Society of Apothecaries of London]

MS e.b. 1 [anonymous recipe collection]

MS l.a. 674 [letter from Mercurius Patten to Walter Bagot, 20 June 1620]

MS v.a. 20 [recipe collection bearing the ownership note of Constance Hall (1672)]

MS v.a. 21 [anonymous recipe collection]

MS v.a. 136 [Thomas Lodge, "The Poor Mans Talent"]

MS v.a. 140 [anonymous recipe collection with personal correspondence]

MS v.a. 215 [recipe collection bearing the ownership note of Susanna Packe (1674)]

MS v.a. 272 [anonymous recipe collection]

MS v.a. 339 [commonplace book of Joseph Hall with medical recipes]

MS v.a. 345 [commonplace book with medical recipes]

MS v.a. 347 [recipe collection with the ownership note of Dorothy Philips (1617); originally a book of sermon notes]

MS v.a. 361 [anonymous recipe collection]

MS v.a. 364 [recipe collection with the ownership notes of Nicholas Webster and Robert Nalson (1670)]

MS v.a. 365 [recipe collection associated with Sir Gilbert Hoghton]

MS v.a. 366 [anonymous recipe collection]

MS v.a. 388 ["A book of such medicines as have been approved by the speciall practize of Mrs Corlyon"]

MS v.a. 396 [recipe collection bearing the ownership note of Penelope Jephson (1674)]

MS v.a. 397 [recipe collection bearing the ownership note of Katherine Brown]

MS v.a. 401 [recipe collection bearing the ownership notes of Jane Staveley (1693) and Henrietta Elizabeth Harrison (1822)]

MS v.a. 422 [the account book of the Hammond family with medical recipes]

MS v.a. 425 [recipe collection bearing the ownership note of Sarah Long]

MS v.a. 429 [recipe collection bearing the ownership notes of Rose Kendall and Ann Cater (1682) and Anna Maria Wentworth (1725/26)]

MS v.a. 430 [recipe collection bearing the ownership notes of Anne Dewes Bradley and Mary Granville and dated 1740]

MS v.a. 450 [recipe collection with the ownership notes of Lettice Pudsey and Elizabeth Jackson]

MS v.a. 452 ["A book of choice receipts," Thomas Sheppey (c. 1675)]

MS v.a. 456 [recipe collection bearing the ownership notes of Mary Baumfylde (1626), Abraham Somers, Katherine Foster (1707), and Catherine Thatcher (1712)]

MS v.a. 458 [anonymous recipe collection]

MS v.a. 468 [recipe collection with the ownership note of Elizabeth Fowler (1684)]

MS v.a. 490 [anonymous recipe collection]

MS v.a. 563 [anonymous recipe collection]

MS v.a. 600 [recipe collection with the ownership note of Lady Grace Castleton]

MS v.b. 13 [anonymous recipe collection]

MS v.b. 14 [recipe collection with the ownership note of Jane Dawson]

MS v.b. 110 [commonplace book of Henry Oxenden with medical recipes]

MS v.b. 129 [recipe collection associated with John Feckenham]

MS v.b. 252 [anonymous recipe collection]

MS v.b. 272 [anonymous recipe collection]

MS v.b. 273 [anonymous recipe collection]

MS v.b. 286 [anonymous recipe collection, including an extract from the collection of the Earl of Chesterfield]

MS w.a. 87 [anonymous recipe collection]

MS w.a. 101 [recipe collection bearing the ownership note of Diana Cunningham]

MS w.a. 111 [anonymous recipe collection]

MS w.a. 317 [anonymous recipe collection]

MS w.b. 456 [recipe collection of Lettis Versey]

MS x.d. 469 [anonymous recipe collection]

Annotated Printed Books

Libraries in the United Kindom

London, British Library

   1038.i.35 (7) & 1038.k.47 (3) [Copy of *A Rich Store-house* (1607), with the ownership note of Matthew Wattes (1610)]

Oxford, Balliol College

   910.i.7 [a collection of nine tracts bought and bound together by Balliol fellow Nicholas Crouch]

Oxford, Bodleian Library

   20 d.6 [Thomas Brugis, *Marrow of Physick* (1640)]

   8° C 571 Linc [Elizabeth Grey, *A Choice Manual* (1653)]

   Douce C 25 (1) [*A Closet for Ladies and Gentlewomen* (1630) with the ownership note of Dorothy Hawkins (1668)]

   Douce P 412 [*The Accomplish'd Ladies Delight* (1675) bearing the ownership note of Narcissus Luttrell (1675)]

   Douce R 66 [*Natura Exenterata* (1655) bearing the ownership note of Mary Mott (June 29 1703)]

   G. Pamph. 2156 (1) [T. C., *A Hospitall for the Diseased* (1579)]

Locke 7.411 [Kenelm Digby, *Choice Experimented Receipts in Physick* (1668) bearing the ownership note of John Locke]

Tanner 212 (2) [Alessio Piemontese, *The Seconde Part of the Secretes of Master Alexis of Piemont* (1566) bearing the ownership notes of Martinus Cobb and Thomas Tanner]

Tanner 212 (3) [Alessio Piemontese, *The Thyrd and Last Parte of the Secretes of the Reverende Mayster Alexis of Piemont* (1566) bearing the ownership notes of Martinus Cobb and Thomas Tanner]

Tanner 880 [*A Hundred and Fourtene Experiments and Cures* (1583) bearing the ownership notes of Robert Cook (1587), Thomas Tanner, and Robert Estrake]

Vet. A1 e.4 [Peter Leven, *A Right and Profitable Booke for All Diseases, Called the Pathway to Health* (1596)]

Vet. A1 e.102 [Thomas Lupton, *A Thousand Notable Things* (1579)]

Vet. A2 f.1 [*A Closet for Ladies and Gentlewomen* (1611)]

Vet. A3 f.1788 [W. M., *The Queens Closet Opened* (1658)]

Vet. A3 f.1872 [Ralph William, *Physical Rarities* (1666) with ownership notes of II. Montgomerie and William Milne (1743)]

Oxford, Radcliffe Science Library

RR z. 179 [Jean Goeurot, *The Regiment of Life* (1560)]

Libraries in the United States

San Marino, California, Henry E. Huntington Library

Rare Book 53930 [Thomas Elyot, *The Castell of Helth* (1572)]

Rare Book 124144 [Thomas Brugis, *Vade Mecum* (1652)]

Rare Book 389398 [Thomas Brugis, *The Marrow of Physick* (1640)]

Washington, DC, Folger Shakespeare Library

11798 copy 3 [Conrad Gesner, *The Newe Jewell of Health* (1576)]

STC 296 copy 2 [Girolamo Ruscelli, *The Secretes of the Reverend Maister Alexis of Piemont* (1562)]

STC 301 copy 2 [Girolamo Ruscelli, *The Secretes of the Reverend Maister Alexis of Piemont* (1562)]

STC 305 copy 2 [Girolamo Ruscelli, *The Secretes of the Reverend Maister Alexis of Piemont* (1562)]

151–446q [Kenelm Digby, *Choice and Experimented Receipts in Physick and Chirurgery* (1668)]

138–469q [*The Queens Closet Opened* (1696) with multiple ownership notes]

Printed Primary Sources

Anselment, Raymond, ed. *The Remembrances of Elizabeth Freke, 1671–1714*, vol. 18. Royal Historical Society Camden, 5th ser. 5. Cambridge: Cambridge University Press, 2001.

Arcana Fairfaxiana Manuscripta: A Manuscript Volume of Apothecaries Lore and Housewifery Nearly Three Centuries Old, Used and Partly Written by the Fairfax Family. With an Introduction by George Wendell. Newcastle-upon-Tyne, UK: Mawson, Swan and Morgan, 1890.

Bate, George, William Salmon, Frederick Hendrick van Hove, and James Shipton. *Pharmacopoeia Bateana.* London: S. Smith and B. Walford, 1700.

Berlu, John Jacob. *The Treasury of Drugs Unlock'd.* London: John Harris and Tho. Howkins, 1690.

Boyle, Robert. "An Invitation to a Free and Generous Communication of Secrets and Receits in Physick." In *Chymical, Medicinal, and Chyrurgical Addresses to Samuel Hartlib.* London, 1655.

———. *Medicinal Experiments, or A Collection of Choice and Safe Remedies for the Most Part Simple and Easily Prepared.* London: Samuel Smith, S. Walford, 1692.

———. *Some Receipts of Medicines: For the Most Part Parable and Simple, Sent to a Friend in America.* London, 1688.

Bradwell, Stephen. *Helps for Suddain Accidents Endangering Life.* London: T. Slater, 1633.

Brugis, Thomas. *The Marrow of Physicke, or A Learned Discourse of the Severall Parts of Mans Body.* London, 1640.

Carne, Brian. *Some St. John Family Papers: Transcribed, with Comments, by Canon Brian Carne.* Reproduced from the *Reports of the Friends of Lydiard Tregoz,* 27–29 (1994–96). Swindon, 1996. https://www.friendsoflydiardpark.org.uk/.

The Country-Man's Apothecary, or A Rule by Which Country Men Safely Walk in Taking Physick; Not Unusefull for Cities. A Treatise Shewing What Herbe, Plant, Root, Seed or Mineral, May Be Used in Physicke in the Room of That Which Is Wanting. London, 1649.

Crooke, Helkiah, and Alexander Read. *Sōmatographia Anthrōpinē.* London, 1616.

Culpeper, Nicholas. *A Physicall Directory, or A Translation of the London Dispensatory Made by the Colledge of Physicians in London.* London, 1649.

Dickinson, Francisco. *A Precious Treasury of Twenty Rare Secrets.* London, 1649.

Digby, Kenelm. *A Choice Collection of Rare Secrets and Experiments in Philosophy.* London, 1682.

———. *Choice and Experimented Receipts in Physick and Chirurgery.* London, 1668.

———. *The Closet of the Eminently Learned Sir Kenelme Digbie Kt. Opened.* London: H. Brome, 1669.

The Experienced Market Man and Woman, or Profitable Instructions to All Masters and Mistrisses of Families, Servants and Others. Edinburgh: James Watson, 1699.

Gabelkover, Oswald. *Artzneybuch.* Tübingen, 1594.

———. *The Boock of Physicke.* Dorte, 1599.

Gerard, John. *The Great Herball, or Generall Historie of Plantes.* London, 1597.

Gethin, Grace. *Misery's Virtues Whetstone: Reliquiae Gethinianae.* London, 1699.

Grey, Elizabeth. *A Choice Manual of Rare and Select Secrets in Physick and Chyrurgery.* London, 1653.

Guybert, Philibert. *Le médecin charitable enseignant la manière de faire et de préparer en la maison avec facilité et peu de frais les remèdes propres à toutes maladies, selon l'advis du médecin ordinaire.* Lyon, 1623.

Harvey, Gideon. *The Family-Physician, and the House-Apothecary.* London: M. Rookes, 1676.

Heresbach, Conrad. *Foure Bookes of Husbandry.* London, 1578.

La Fountaine, Edward. *A Brief Collection of Many Rare Secrets.* London, 1650.

Lovell, William. *The Dukes Desk Newly Broken Up.* 1660.

M., A. *Queen Elizabeths Closset of Physical Secrets.* London: Will. Sheares Junior, 1652.

M., W. *The Queens Closet Opened.* London: Nathaniel Brooks, 1655.

Markham, Gervase. *The English Hus-wife.* London: Roger Jackson, 1615.

Palmer, Archdale. *The Recipe Book 1659–1672 of Archdale Palmer, Gent. Lord of the Manor of Wanlip in the County of Leicestershire*, edited by B. G. Grant Uden. Wymondham, UK: Sycamore Press, 1985.

Perkins, Francis. *Perkins. A New Almanack . . .* London: Company of Stationers, 1730.

Quincy, John. *Compleat English Dispensatory.* London, 1718.

Read, Alexander. *Most Excellent and Approved Medicines and Remedies for Most Diseases and Maladies Incident to Man's Body.* London: George Latham Junior, 1652.

———. *A Treatise of the First Part of Chirurgerie, Called by Mee Synthetikē, the Part Which Teacheth the Reunition of the Parts of the Bodie Disjoyned.* London: Francis Constable, 1638.

Scappi, Bartolomeo. *Opera dell'arte del cucinare.* Venice, 1570.

Smith, John. *The Generall Historie of Virginia, New-England, and the Summer Isles.* London: Printed by I[ohn] D[awson] and I[ohn] H[aviland] for Michael Sparkes, 1626.

T., A. *A Rich Store-house or Treasury for the Diseased.* London: Thomas Purfoot, Raph Blower, 1596.

Tryon, Thomas. *A New Art of Brewing Beer, Ale, and Other Sorts of Liquors.* London: T. Salusbury, 1690.

White, John. *A Rich Cabinet, with Variety of Inventions.* London, 1689.

Winter, Salvator, and Francisco Dickinson. *A Pretious Treasury, or A New Dispensatory.* London: Richard Harper, 1649.

Wood, Owen. *Choice and Profitable Secrets both Physical and Chirurgical: Formerly Concealed by the Deceased Dutchess of Lenox*. London, 1656.

———. *An Epitomie of Most Experienced, Excellent and Profitable Secrets appertaining to Physick and Chirurgery*. London, 1651.

Wood, Owen, and Alexander Read. *An Alphabetical Book of Physicall Secrets, for All Those Diseases That Are Most Predominant and Dangerous (Curable by Art) in the Body of Man*. London, 1639.

Woolley, Hannah. *The Compleat Servant-Maid, or The Young Maidens Tutor*. London, 1677.

———. *The Cook's Guide, or Rare Receipts for Cookery*. London, 1664.

Yworth, William. *Cerevisiarii Comes, or The New and True Art of Brewing Beer, Ale and Other Liquors*. London: J. Taylor and S. Clement, 1692.

Secondary Sources

Adams, J. N., and Marilyn Deegan. "Bald's *Leechbook* and the *Physica Plinii*." *Anglo-Saxon England* 21 (1992): 87–114.

Adams, Thomas R., and Nicolas Barker. "A New Model for the Study of the Book." In *A Potencie of Life: Books in Society; The Clark Lectures 1986–1987*, 5–43. London: British Library, 1993.

Agrimi, Jole, and Chiara Crisciani. "Per una ricerca su experimentum-experimenta: Riflessione epistemologica e tradizione medica (secoli XIII–XV)." In *Presenza del lessico greco et latino nelle linque contemporanee*, edited by Pietro Janni and Innocenzo Mazzini, 9–49. Macerata, Italy: Facoltà di Lettere e Filosofia, Università degli Studi di Macerata, 1990.

Albala, Ken. *Eating Right in the Renaissance*. Berkeley: University of California Press, 2002.

Algazi, Gadi. "Scholars in Households: Refiguring the Learned Habitus, 1480–1550." *Science in Context* 16, nos. 1–2 (2003): 9–42.

Alonso-Almeida, Francisco, and Mercedes Cabrera-Abreu. "The Formulation of Promise in Medieval English Medical Recipes: A Relevance-Theoretic Approach." *Neophilologus* 86, no. 1 (2002): 137–54.

Andersen, Jennifer, and Elizabeth Sauer, eds. *Books and Readers in Early Modern England: Material Studies*. Philadelphia: University of Pennsylvania Press, 2002.

Andrews, Jonathan. "Napier, Richard (1559–1634)." In *Oxford Dictionary of National Biography*. Oxford: Oxford University Press, 2004–. Accessed 19 October 2017. http://www.oxforddnb.com/view/article/19763.

Anselment, Raymond A. "Lovelace, Richard (1617–1657), poet and army officer." In *Oxford Dictionary of National Biography*. Oxford: Oxford University Press, 2004–. Accessed 4 February 2018. http://www.oxforddnb.com/view/10.1093/ref:odnb/9780198614128.001.0001/odnb-9780198614128-e-17056.

Archer, Ian. "Social Networks in Restoration London: The Evidence from Samuel Pepys's Diary." In *Communities in Early Modern England*, edited by Alexandra Shepard and Phil Withington, 76–94. Manchester: Manchester University Press, 2000.

Archer, Jayne. "The Queens' Arcanum: Authority and Authorship in the *Queens Closet Opened* (1655)." *Renaissance Journal* 1, no. 6 (2002): 14–26.

Ash, Eric H. *Power, Knowledge, and Expertise in Elizabethan England*. Baltimore: Johns Hopkins University Press, 2004.

Aspin, Richard. "Who Was Elizabeth Okeover?" *Medical History* 44, no. 4 (2000): 531–40.

Aughterson, Kate. "Brockman, Ann, Lady Brockman (*d.* 1660)." In *Oxford Dictionary of National Biography*. Oxford: Oxford University Press, 2004–. Accessed 19 October 2017. http://www.oxforddnb.com/view/article/68030.

Baker, Joanne. "Female Monasticism and Family Strategy: The Guises and Saint Pierre de Reims." *Sixteenth Century Journal* 28 (1997): 1091–108.

Barker, Sheila. "Christine de Lorraine and Medicine at the Medici Court." In *Medici Women: The Making of a Dynasty in Grand Ducal Tuscany*, edited by Judith C. Brown and Giovanna Benadusi, 155–81. Toronto: Centre for Reformation and Renaissance Studies, 2015.

Bassnett, Madeline. "Restoring the Royal Household: Royalist Politics and the Commonwealth Recipe Book." *Early English Studies* 2 (2009): 1–32.

Beier, Lucinda McCray. "In Sickness and in Health: A Seventeenth Century Family's Experience." In *Patients and Practitioners: Lay Perceptions of Medicine in Pre-Industrial Society*, edited by Roy Porter, 101–28. Cambridge: Cambridge University Press, 1985.

———. *Sufferers and Healers: The Experience of Illness in Seventeenth-Century England*. London: Routledge and Kegan Paul, 1987.

Bell, H. I. "The Brockman Charters." *British Museum Quarterly* 6, no. 3 (1931): 75.

Ben-Amos, Ilana Krausman. *The Culture of Giving: Informal Support and Gift-Exchange in Early Modern England*. Cambridge: Cambridge University Press, 2008.

Birken, William. "Bathurst, John (d. 1659), physician." In *Oxford Dictionary of National Biography*. Oxford: Oxford University Press, 2004–. Accessed 31 January 2018. http://www.oxforddnb.com/view/10.1093/ref:odnb/9780198614128.001.0001/odnb-9780198614128-e-1698.

———. "Micklethwaite, Sir John (*bap.* 1612, *d.* 1682)." In *Oxford Dictionary of National Biography*. Oxford: Oxford University Press, 2004–. Accessed 19 October 2017. http://www.oxforddnb.com/view/article/18662.

Blair, Ann. *Too Much to Know: Managing Scholarly Information before the Modern Age*. New Haven, CT: Yale University Press Press, 2010.

Bray, Francesca. "Gender and Technology." *Annual Review of Anthropology* 36 (2007): 37–53.

———. *Technology, Gender and History in Imperial China: Great Transformations Reconsidered*. Abingdon, UK: Routledge, 2013.

Brockliss, Laurence, and Colin Jones. *The Medical World of Early Modern France*. Oxford: Oxford University Press, 1997.

Broomhall, Susan. *Women's Medical Work in Early Modern France*. Manchester: Manchester University Press, 2011.

Brown, Nancy Pollard. "Howard [Dacre], Anne, Countess of Arundel (1557–1630)." In *Oxford Dictionary of National Biography*. Oxford: Oxford University Press, 2004–. Accessed 19 October 2017. http://www.oxforddnb.com/view/article/46907.

Brown, P. S. "The Venders of Medicines Advertised in Eighteenth-Century Bath Newspapers." *Medical History* 19, no. 4 (1975): 352–69.

Burke, Peter. *What Is the History of Knowledge?* Cambridge: Polity, 2015.

Burke, Victoria. "Ann Bowyer's Commonplace Book (Bodleian Library Ashmole MS 51): Reading and Writing among the 'Middling Sort.'" *Early Modern Literary Studies* 6, no. 3 (2001): 1–28.

———. "Recent Studies in Commonplace Books." *English Literary Renaissance* 43, no. 1 (2013): 153–77.

Burns, Kathryn. *Into the Archive: Writing and Power in Colonial Peru*. Durham, NC: Duke University Press, 2010.

Bushnell, Rebecca. "The Gardener and the Book." In *Didactic Literature in England, 1500–1800: Expertise Construed*, edited by Natasha Glaisyer and Sara Pennell, 118–36. Aldershot, UK: Ashgate, 2003.

———. *Green Desire: Imagining Early Modern English Gardens*. Ithaca, NY: Cornell University Press, 2003.

Butcher, Burford. "The Brockman Papers." *Archaeologia Cantiana* 43 (1931): 281–83.

Bycroft, Michael. "Iatrochemistry and the Evaluation of Mineral Waters in France, 1600–1750." *Bulletin of the History of Medicine* 91, no. 2 (2017): 303–30.

Cabré, Montserrat. "Women Healers? Household Practices and the Categories of Health Care in Late Medieval Iberia." *Bulletin of the History of Medicine* 82, no. 1 (2008): 18–51.

Cameron, M. L. "Bald's *Leechbook*: Its Sources and Their Use in Its Compilation." *Anglo-Saxon England* 12 (1983): 153–82.

———. "The Sources of Medical Knowledge in Anglo-Saxon England." *Anglo-Saxon England* 11 (1982): 135–55.

Capp, Bernard. *When Gossips Meet: Women, Family, and Neighbourhood in Early Modern England*. Oxford: Oxford University Press, 2003.

Cavallo, Sandra. *Artisans of the Body in Early Modern Italy: Identities, Families and Masculinities*. Manchester: Manchester University Press, 2007.

Cavallo, Sandra, and Tessa Storey. *Healthy Living in Late Renaissance Italy*. Oxford: Oxford University Press, 2013.

Certeau, Michel de. *The Practice of Everyday Life*. Translated by Steven Rendall. Berkeley: University of California Press, 1984.

Charmantier, Isabelle, and Stefan Müller-Wille, eds. "Worlds of Paper." *Early Science and Medicine* 19 (2014).

Chartier, Roger. *The Order of Books: Readers, Authors, and Libraries in Europe between the 14th and 18th Centuries.* Stanford, CA: Stanford University Press, 1994.

Cicchetti, Angelo, and Raul Mordenti. *I libri di famiglia in Italia. Vol. 1, Filologia e storiografia letteraria.* Rome: Edizioni di Storia e Letteratura, 1985.

Clark, Alice. *Working Life of Women in the Seventeenth Century.* 1919. Reprint, London: Routledge, 1992.

Clay, J. W. *Dugdale's Visitation of Yorkshire with Additions.* Exeter, 1901.

Considine, John. "Wolley, Hannah (*b.* 1622?, *d.* in or after 1674)." In *Oxford Dictionary of National Biography.* Oxford: Oxford University Press, 2004. Accessed 19 October 2017. http://www.oxforddnb.com/view/article/29957.

Cook, Harold J. *The Decline of the Old Medical Regime in Stuart London.* Ithaca, NY: Cornell University Press, 1986.

———. "The History of Medicine and the Scientific Revolution." *Isis* 102, no. 1 (2011): 102–8.

———. *Matters of Exchange: Commerce, Medicine, and Science in the Dutch Golden Age.* New Haven, CT: Yale University Press, 2007.

———. "Sir John Colbatch and Augustan Medicine: Experimentalism, Character and Entrepreneurialism." *Annals of Science* 47, no. 5 (1990): 475–505.

Cooper, Alix. "Homes and Households." In *Early Modern Science,* edited by Katharine Park and Lorraine Daston, vol. 3 of *The Cambridge History of Science,* 224–37. Cambridge: Cambridge University Press, 2006.

———. "Picturing Nature: Gender and the Politics of Natural-Historical Description in Eighteenth-Century Gdańsk/Danzig." *Journal for Eighteenth-Century Studies* 36, no. 4 (2013): 519–29.

Crawford, Patricia. "Women's Published Writings, 1600–1700." In *Women in English Society, 1500–1800,* edited by Mary Prior, 211–82. 1985. Reprint, London: Routledge, 1996.

Cust, Richard. "Grosvenor, Sir Richard, first baronet (1585–1645), magistrate and politician." In *Oxford Dictionary of National Biography.* Oxford: Oxford University Press, 2004–. Accessed 4 February 2018. https://doi.org/10.1093/ref:odnb/37492.

Darnton, Robert. "First Steps toward a History of Reading." *Australian Journal of French Studies* 23, no. 1 (1986): 5–30.

———. "What Is the History of Books?" In *The Kiss of Lamourette: Reflections in Cultural History,* edited by Robert Darnton, 107–36. London: W. W. Norton, 1991.

———. "'What Is the History of Books?' Revisited." *Modern Intellectual History* 4 (2007): 495–508.

Daston, Lorraine. "The History of Science and the History of Knowledge." *KNOW: A Journal on the Formation of Knowledge* 1, no. 1 (2017): 131–54.

———. "The Sciences of the Archive." *Osiris* 27, no. 1 (2012): 156–87.

———. "Taking Note(s)." *Isis* 95, no. 3 (2004): 443–48.

Daston, Lorraine, and Elizabeth Lunbeck. *Histories of Scientific Observation*. Chicago: University of Chicago Press, 2011.

Daston, Lorraine, and Katharine Park. *Wonders and the Order of Nature, 1150–1750*. Cambridge, MA: Zone Books, 2001.

Davis, Natalie Zemon. "Beyond the Market: Books as Gifts in Sixteenth-Century France." *Transactions of the Royal Historical Society*, 5th ser., 33 (1983): 69–88.

———. "Ghosts, Kin, and Progeny: Some Features of Family Life in Early Modern France." *Daedalus* 106, no. 2 (1977): 87–114.

———. *The Gift in Sixteenth-Century France*. Madison: University of Wisconsin Press, 2000.

Dawson, Warren R., ed. *A Leechbook or Collection of Medical Recipes of the Fifteenth Century*. London: Macmillan, 1934.

Daybell, James. *The Material Letter in Early Modern England: Manuscript Letters and the Culture and Practices of Letter-Writing, 1512–1635*. London: Palgrave Macmillan, 2012.

Dear, Peter. "The Meanings of Experience." In *Early Modern Science*, edited by Katharine Park and Lorraine Daston, vol. 3 of *The Cambridge History of Science*, 106–31. Cambridge: Cambridge University Press, 2006.

———. "*Totius in Verba*: Rhetoric and Authority in the Early Royal Society." *Isis* 76, no. 2 (1985): 144–61.

DiMeo, Michelle. "Authorship and Medical Networks: Reading Attributions in Early Modern Manuscript Recipe Books." In *Reading and Writing Recipe Books, 1550–1800*, edited by Michelle DiMeo and Sara Pennell, 25–46. Manchester: Manchester University Press, 2013.

———. "'Communicating Medical Recipes': Robert Boyle's Genre and Rhetorical Strategies for Print." In *The Palgrave Handbook of Early Modern Literature and Science*, edited by Howard Marchitello and Evelyn Tribble, 209–28. Basingstoke, UK: Palgrave Macmillan, 2016.

DiMeo, Michelle, and Sara Pennell. "Introduction." In *Reading and Writing Recipe Books, 1550–1800*, edited by Michelle DiMeo and Sara Pennell, 1–24. Manchester: Manchester University Press, 2013.

———, eds. *Reading and Writing Recipe Books, 1550–1800*. Manchester: Manchester University Press, 2013.

Doherty, Francis. "The Anodyne Necklace: A Quack Remedy and Its Promotion." *Medical History* 34, no. 3 (1990): 268–93.

Donawerth, Jane. "Women's Poetry and the Tudor-Stuart System of Gift Exchange." In *Women, Writing and the Reproduction of Culture in Tudor and Stuart Britain*, edited by Mary E. Burke, Jane Donawerth, Linda L. Dove, and Karen Nelson, 3–18. Syracuse, NY: Syracuse University Press, 2000.

Drake-Brockman, Giles. "Sir William and Lady Ann Brockman of Beachborough,

Newington by Hythe: A Royalist Family's Experience of the Civil War." *Archaeologia Cantiana* 132 (2012): 21–41.

Dunning, R. W. "Brewham." In *Bruton, Horethorne and Norton Ferris Hundreds*, vol. 7 of *A History of the County of Somerset*. Oxford: Oxford University Press, 1999.

Dunning, R. W., K. H. Rogers, P. A. Spalding, Colin Shrimpton, Janet H. Stevenson, and Margaret Tomlinson. "Parishes: Lydiard Tregoze." In *A History of the County of Wiltshire*, vol. 9, edited by Elizabeth Crittall, 75–90. London: Victoria County History, 1970.

Eamon, William. *The Professor of Secrets: Mystery, Medicine, and Alchemy in Renaissance Italy*. Washington, DC: National Geographic Society, 2010.

———. *Science and the Secrets of Nature: Books of Secrets in Medieval and Early Modern Culture*. Princeton, NJ: Princeton University Press, 1994.

Edgerton, David. *The Shock of the Old: Technology and Global History since 1900*. London: Profile Books, 2006.

Elzinga, J. G. "Howard, Philip [St. Philip Howard], thirteenth Earl of Arundel (1557–1595)." In *Oxford Dictionary of National Biography*. Oxford: Oxford University Press, 2004–. Accessed 19 October 2017. http://www.oxforddnb.com/view/article/13929.

Erickson, Amy Louise. *Women and Property in Early Modern England*. Abingdon, UK: Routledge, 1993.

Evans, Robert, John Weston, and Alexander Marr. *Curiosity and Wonder from the Renaissance to the Enlightenment*. Aldershot, UK: Ashgate, 2006.

Evenden, Doreen. *Popular Medicine in Seventeenth-Century England*. Bowling Green, OH: Bowling Green State University Popular Press, 1988.

Ezell, Margaret J. M. "Cooking the Books, or The Three Faces of Hannah Woolley." In *Reading and Writing Recipe Books, 1550–1800*, edited by Michelle DiMeo and Sara Pennell, 159–78. Manchester: Manchester University Press, 2013.

———. "Domestic Papers: Manuscript Culture and Early Modern Women's Life Writing." In *Genre and Women's Life Writing in Early Modern England*, edited by Michelle M. Dowd and Julie A. Eckerle, 33–48. Aldershot, UK: Ashgate, 2007.

———. "Invisible Books." In *Producing the Eighteenth-Century Book: Writers and Publishers in England, 1650–1800*, edited by Pat Rogers and Laura Runge, 53–69. Newark: University of Delaware Press, 2009.

———. *Social Authorship and the Advent of Print*. Baltimore: Johns Hopkins University Press, 2003.

Ferguson, John K. *Bibliographical Notes on Histories of Inventions and Books of Secrets*. 2 vols. 1898. Reprint, London: Holland Press, 1959.

Field, Catherine. "Kitchen Science and the Public/Private Role of the Housewife Practitioner." Paper presented at the Women on the Verge of Science seminar at the Folger Institute, 14 May 2003.

———. "'Many Hands Hands': Writing the Self in Early Modern Women's Recipe Books." In *Genre and Women's Life Writing in Early Modern England*, edited by Michelle M. Dowd and Julie A. Eckerle, 49–64. Aldershot, UK: Ashgate, 2007.

Findlen, Paula. "The Economy of Scientific Exchange in Early Modern Italy." In *Patronage and Institutions: Science, Technology, and Medicine at the European Court, 1500–1750*, edited by Bruce T. Moran, 5–24. Rochester, NY: Boydell Press, 1991.

———. "Masculine Prerogatives: Gender, Space and Knowledge in the Early Modern Museum." In *The Architecture of Science*, edited by Peter Galison and Emily Thompson, 29–58. Cambridge, MA: MIT Press, 1999.

Fissell, Mary. "The Marketplace of Print." In *Medicine and the Market in England and Its Colonies, c. 1450–c. 1850*, edited by Mark Jenner and Patrick Wallis, 108–32. Basingstoke, UK: Palgrave Macmillan, 2007.

———. "Popular Medical Writing." In *Cheap Print in Britain and Ireland to 1660*, vol. 1 of *The Oxford History of Popular Print Culture*, edited by Joad Raymond, 417–30. Oxford: Oxford University Press, 2011.

———. *Vernacular Bodies: The Politics of Reproduction in Early Modern England*. Oxford: Oxford University Press, 2004.

———. "Women, Health and Healing in Early Modern Europe." *Bulletin for the History of Medicine* 82 (2008): 1–17.

Flather, Amanda. *Gender and Space in Early Modern England*. Rochester, NY: Boydell Press, 2011.

Floyd-Wilson, Mary. *Occult Knowledge, Science, and Gender on the Shakespearean Stage*. Cambridge: Cambridge University Press, 2013.

Foster, Donald W. "Stuart, Frances, Duchess of Lennox and Richmond [*other married name* Frances Seymour, Countess of Hertford] (1578–1639)." In *Oxford Dictionary of National Biography*. Oxford: Oxford University Press, 2004–. Accessed 19 October 2017. http://www.oxforddnb.com/view/article/70952.

Foster, Joseph. *Alumni Oxonienses, 1500–1714*. Vol. 1. Oxford: Oxford University Press, 1891.

———. *Alumni Oxonienses: The Members of the University of Oxford, 1715–1886; Their Parentage, Birthplace and Year of Birth, with a Record of Their Degrees. Being the Matriculation Register of the University*. Vol. 1. Oxford: University of Oxford, 1888.

Fox, Adam. *Oral and Literate Culture in England, 1500–1700*. Oxford: Clarendon Press, 2000.

Freke, Ralph, John Freke, and William Freke. *A Pedigree, or Genealogye of the Family of the Freke's, Begun by R. Freke, Augmented by J. Freke, Reduced to This Form by W. Freke, July 14th 1707*. London: Typis Medio-Montanis, 1825.

Furdell, Elizabeth Lane. "Bate, George (1608–1668)." In *Oxford Dictionary of National Biography*. Oxford University Press, 2004–. Accessed 19 October, 2017. http://www.oxforddnb.com/view/article/1661.

————. *Publishing and Medicine in Early Modern England*. Rochester, NY: University of Rochester Press, 2002.

Getz, Faye M. *Healing and Society in Medieval England: A Middle English Translation of the Pharmaceutical Writings of Gilbertus Anglicus*. Madison: University of Wisconsin Press, 1991.

Gevitz, N. "'Helps for Suddain Accidents': Stephen Bradwell and the Origin of the First Aid Guide." *Bulletin of the History of Medicine* 67, no. 1 (1993): 51–73.

Goldie, Mark. "Tyrrell, James (1642–1718), political theorist and historian." In *Oxford Dictionary of National Biography*. Oxford: Oxford University Press, 2004–. Accessed 5 February 2018. https://doi.org/10.1093/ref:odnb/27953.

Goldsmith, Valentine Fernande. *A Short Title Catalogue of French Books, 1601–1700, in the Library of the British Museum*. Folkstone, UK: Dawsons, 1973.

Goodwin, Gordon. "Harley, Sir Edward (1624–1700)." In *Oxford Dictionary of National Biography*. Oxford: Oxford University Press, 2004–. Accessed 19 October 2017. https://doi.org/10.1093/ref:odnb/12335.

Grafton, Anthony, and Nancy G. Siraisi. *Natural Particulars: Nature and the Disciplines in Renaissance Europe*. Cambridge, MA: MIT Press, 1999.

Green, Everard. "Pedigree of Johnson of Ayscough-Fee Hall, Spalding, Co. Lincoln." *Genealogist* 1 (1877): 110.

Green, Monica. *Making Women's Medicine Masculine: The Rise of Male Authority in Pre-Modern Gynecology*. Oxford: Oxford University Press, 2008.

Greengrass, Mark, Michael Leslie, and M. Hannon. *The Hartlib Papers*. Sheffield, UK: HRI Online Publications. https://www.hrionline.ac.uk/hartlib.

Greengrass, Mark, Michael Leslie, and Timothy Raylor, eds. *Samuel Hartlib and Universal Reformation: Studies in Intellectual Communication*. Cambridge: Cambridge University Press, 1994.

Griffin, Andrew. "Banister, John (1532/3–1599?)." In *Oxford Dictionary of National Biography*. Oxford: Oxford University Press, 2004 . Accessed 19 October 2017. http://www.oxforddnb.com/view/article/1280.

Hackel, Heidi Brayman. *Reading Material in Early Modern England: Print, Gender, and Literacy*. Cambridge: Cambridge University Press, 2005.

Hainsworth, D. R. *Stewards, Lords and People: The Estate Steward and His World in Later Stuart England*. Cambridge: Cambridge University Press, 2008.

Handley, S. "Sir Robert Cotton, 1st Bt. (c. 1635–1712), of Combermere, Cheshire." In *The History of Parliament: The House of Commons, 1690–1715*, edited by D. Hayon, E. Cruickshanks, and S. Handley (2002). Accessed 29 January 2018. http://www.historyofparliamentonline.org/volume/1690-1715/member/cotton-sir-robert-1635-1712.

Hanson, Marta, and Gianna Pomata. "Medicinal Formulas and Experiential Knowledge in the Seventeenth-Century Epistemic Exchange between China and Europe." *Isis* 108, no. 1 (2017): 1–25.

Harkness, Deborah E. *The Jewel House: Elizabethan London and the Scientific Revolution*. New Haven, CT: Yale University Press, 2007.

———. "Managing an Experimental Household: The Dees of Mortlake and the Practice of Natural Philosophy." *Isis* 88, no. 2 (1997): 247–62.

Harvey, Karen. *The Little Republic: Masculinity and Domestic Authority in Eighteenth-Century Britain*. Oxford: Oxford University Press, 2012.

Hasted, Edward. "Parishes: Newington." In *The History and Topographical Survey of the County of Kent*, 8:197–210. Canterbury: W. Bristow, 1799. http://www.british-history.ac.uk/vch/surrey/vol4/pp74-77.

Haycock, David Boyd. "Johnson, Maurice (1688–1755)." In *Oxford Dictionary of National Biography*. Oxford: Oxford University Press, 2004–. Accessed 19 October 2017. http://www.oxforddnb.com/view/article/14908.

Heal, Felicity. "Food Gifts, the Household and the Politics of Exchange in Early Modern England." *Past and Present* 199 (2008). 41–70.

———. *Hospitality in Early Modern England*. Oxford: Clarendon Press, 1990.

———. *The Power of Gifts: Gift Exchange in Early Modern England*. Oxford: Oxford University Press, 2014.

Heesen, Anke te. "The Notebook: A Paper-Technology." In *Making Things Public: Atmospheres of Democracy*, edited by Bruno Latour and Peter Weibel, 582–89. Cambridge, MA: MIT Press, 2005.

Heller, Jennifer. *The Mother's Legacy in Early Modern England*. Farnham, UK: Ashgate, 2011.

Herbert, Amanda E. *Female Alliances: Gender, Identity, and Friendship in Early Modern Britain*. New Haven, CT: Yale University Press, 2014.

Hess, Volker, and Andrew Mendelsohn, eds. "Paper Technology in der Frühen Neuzeit." *NTM Zeitschrift für Geschichte der Wissenschaften, Technik und Medizin* 21, no. 1 (2013).

Hindle, Steve. "Below Stairs at Arbury Hall: Sir Richard Newdigate and His Household Staff, c. 1670–1710." *Historical Research* 85, no. 227 (2012): 71–88.

Hobby, Elaine. *Virtue of Necessity: English Women's Writing, 1646–1688*. London: Virago, 1988.

———. "'A Woman's Best Setting Out Is Silence': The Writings of Hannah Wolley." In *Culture and Society in the Stuart Restoration: Literature, Drama, History*, edited by Gerald Maclean, 179–200. Cambridge: Cambridge University Press, 1995.

Hodgkin, Katharine. "Women, Memory and Family History in Seventeenth-Century England." In *Memory before Modernity: Practices of Memory in Early Modern Europe*, edited by Erika Kuijpers, Judith Pollmann, Johannes Müller, and Jasper van der Steen, 297–314. Leiden: Brill, 2013.

Holmes, Frederic L. "Laboratory Notebooks and Investigative Pathways." In *Reworking the Bench: Research Notebooks in the History of Science*, edited by Frederic L. Holmes, Jürgen Renn, and Hans-Jörg Rheinberger, 295–308. Dordrecht: Kluwer, 2003.

Holmes, Frederic L., and Trevor H. Levere, eds. *Instruments and Experimentation in the History of Chemistry.* Cambridge, MA: MIT Press, 2000.

Holmes, Frederic L., Jürgen Renn, and Hans-Jörg Rheinberger, eds. *Reworking the Bench: Research Notebooks in the History of Science.* Dordrecht: Kluwer, 2003.

Hopper, Andrew J. "Fairfax, Ferdinando, second Lord Fairfax of Cameron (1584–1648)." In *Oxford Dictionary of National Biography.* Oxford: Oxford University Press, 2004–. Accessed 19 October 2017. http://www.oxforddnb.com/view/article/9081.

———. "Fairfax, Henry (1588–1665)." In *Oxford Dictionary of National Biography.* Oxford: Oxford University Press, 2004–. Accessed 19 October 2017. http://www.oxforddnb.com/view/article/9084.

Houghton, Walter E. "The History of Trades: Its Relation to Seventeenth-Century Thought; As Seen in Bacon, Petty, Evelyn, and Boyle." *Journal of the History of Ideas* 2, no. 1 (1941): 33–60.

Hunt, Tony. *Popular Medicine in Thirteenth-Century England: Introduction and Texts.* Cambridge: Brewer, 1990.

Hunter, Lynette. "Sisters of the Royal Society: The Circle of Katherine Jones, Lady Ranelagh." In *Women, Science and Medicine, 1500–1700: Mothers and Sisters of the Royal Society*, edited by Lynette Hunter and Sarah Hutton, 178–97. Thrupp, Stroud, Gloucestershire, UK: Sutton, 1997.

———. "Women and Domestic Medicine: Lady Experimenters, 1570–1620." In *Women, Science and Medicine, 1500–1700: Mothers and Sisters of the Royal Society*, edited by Lynette Hunter and Sarah Hutton, 89–107. Thrupp, Stroud, Gloucestershire, UK: Sutton, 1997.

Hunter, Lynette, and Sarah Hutton, eds. *Women, Science and Medicine, 1500–1700: Mothers and Sisters of the Royal Society.* Thrupp, Stroud, Gloucestershire, UK: Sutton, 1997.

Hunter, Michael, ed. *Archives of the Scientific Revolution: The Formation and Exchange of Ideas in Seventeenth-Century Europe.* Woodbridge, UK: Boydell Press, 1998.

———. "The Reluctant Philanthropist: Robert Boyle and the 'Communication of Secrets and Receits in Physick.'" In *Religio Medici: Medicine and Religion in Seventeenth-Century England*, edited by Ole Peter Grell and Andrew Cunningham, 247–72. Aldershot, UK: Ashgate, 1996.

———. *Science and Society in Restoration England.* Cambridge: Cambridge University Press, 1981.

Jackson, H. J. *Marginalia: Readers Writing in Books.* New Haven, CT: Yale University Press, 2002.

Jardine, Lisa, and Anthony Grafton. "'Studied for Action': How Gabriel Harvey Read His Livy." *Past and Present* 129, no. 1 (1990): 30–78.

Jardine, Lisa, and William Sherman. "Pragmatic Readers: Knowledge Transactions and Scholarly Services in Late Elizabethan England." In *Religion, Culture and*

Society in Early Modern Britain, edited by Anthony Fletcher and Peter Roberts, 102–24. Cambridge: Cambridge University Press, 1994.

Johns, Adrian. *The Nature of the Book: Print and Knowledge in the Making*. Chicago: University of Chicago Press, 1998.

Johnston, Dorothy. "The Life and Domestic Context of Francis Willughby." In *Virtuoso by Nature: The Scientific Worlds of Francis Willughby FRS (1635–1672)*, edited by Tim Birkhead, 1–43. Leiden: Brill, 2016.

Jones, Claire. "Formula and Formulation: 'Efficacy Phrases' in Medieval English Medical Manuscripts." *Neuphilologische Mitteilungen* 99 (1998): 199–209.

Jones, Peter. "The *Tabula Medicine*: An Evolving Encyclopaedia." In *English Manuscript Studies, 1100–1700*, vol. 14, *Regional Manuscripts, 1200–1700*, edited by A. S. G. Edwards, 60–85. London: British Library, 2008.

Kassell, Lauren. "Casebooks in Early Modern England: Medicine, Astrology, and Written Records." *Bulletin of the History of Medicine* 88, no. 4 (2014): 595–625.

———. *Medicine and Magic in Elizabethan London: Simon Forman, Astrologer, Alchemist, and Physician*. Oxford: Oxford University Press, 2005.

———. "Secrets Revealed: Alchemical Books in Early Modern England." *History of Science* 49, no. 1 (2011): 61–125.

Kavey, Allison. *Books of Secrets: Natural Philosophy in England, 1550–1600*. Urbana: University of Illiniois Press, 2007.

Ketelaar, Eric. "The Genealogical Gaze: Family Identities and Family Archives in the Fourteenth to Seventeenth Centuries." *Libraries and the Cultural Record* 44, no. 1 (2009): 9–28.

———. "Muniments and Monuments: The Dawn of Archives as Cultural Patrimony." *Archival Science* 7, no. 4 (2007): 343–57.

King, Steven, and Alan Weaver. "Lives in Many Hands: The Medical Landscape in Lancashire, 1700–1820." *Medical History* 44, no. 2 (2000): 173–200.

Klein, Ursula, and Wolfgang Lefèvre. *Materials in Eighteenth-Century Science: A Historical Ontology*. Cambridge, MA: MIT Press, 2007.

Klein, Ursula, and Emma C. Spary, eds. *Materials and Expertise in Early Modern Europe: Between Market and Laboratory*. Chicago: University of Chicago Press, 2010.

Knight, E. B. *The Visitation of Oxfordshire, 1669 and 1675*. London: Harleian Society, 1993.

Knight, Leah. *Of Books and Botany in Early Modern England*. Farnham, UK: Ashgate, 2009.

Knoppers, Laura L. "Opening the Queen's Closet: Henrietta Maria, Elizabeth Cromwell, and the Politics of Cookery." *Renaissance Quarterly* 60, no. 2 (2007): 464–99.

Knowles, James. "Conway, Edward, second Viscount Conway and Second Viscount Killultagh (*bap.* 1594, *d.* 1655)." In *Oxford Dictionary of National Biography*. Oxford: Oxford University Press, 2004–. Accessed 19 October 2017. http://www.oxforddnb.com/view/article/55441.

Kowalchuk, Kristine. *Preserving on Paper: Seventeenth-Century Englishwomen's Receipt Books*. Toronto: University of Toronto Press, Scholarly Publishing Division, 2017.

Krämer, Fabian. "Ulisse Aldrovandi's *Pandechion Epistemonicon* and the Use of Paper Technology in Renaissance Natural History." *Early Science and Medicine* 19 (2014): 398–423.

Krämer, Fabian, and Helmut Zedelmaier. "Instruments of Invention in Renaissance Europe: The Cases of Conrad Gesner and Ulisse Aldrovandi." *Intellectual History Review* 24, no. 3 (2014): 321–41.

Krohn, Deborah L. *Food and Knowledge in Renaissance Italy: Bartolomeo Scappi's Paper Kitchens*. Farnham, UK: Ashgate, 2015.

Laroche, Rebecca. *Medical Authority and Englishwomen's Herbal Texts, 1550–1650*. Aldershot, UK: Ashgate, 2009.

Laroche, Rebecca, and Steven Turner. "Robert Boyle, Hannah Woolley, and Syrup of Violets." *Notes and Queries* 58 (2011): 390–91.

Lehmann, Gilly. *The British Housewife: Cookery Books, Cooking and Society in Eighteenth-Century Britain*. Totnes, Devon, UK: Prospect Books, 2003.

LeJacq, Seth. "The Bounds of Domestic Healing: Medical Recipes, Storytelling and Surgery in Early Modern England." *Social History of Medicine* 26, no. 3 (2013): 451–68.

Leong, Elaine. "Brewing Ale and Boiling Water in 1651." In *Structures of Practical Knowledge*, edited by Matteo Valleriani, 55–75. Heidelberg: Springer, 2017.

———. "Collecting Knowledge for the Family: Recipes, Gender and Practical Knowledge in the Early Modern English Household." *Centaurus* 55, no. 2 (2013): 81–103.

———. "'Herbals She Peruseth': Reading Medicine in Early Modern England." *Renaissance Studies* 28, no. 4 (2014): 556–78.

———. "Making Medicines in the Early Modern Household." *Bulletin of the History of Medicine* 82, no. 1 (2008): 145–68.

———. "Medical Recipe Collections in Seventeenth-Century England: Knowledge, Text and Gender." PhD diss., University of Oxford, 2006.

———. "Printing Vernacular Medicine in Early Modern England: The Case of Lazare Rivière's *The Practice of Physick*." In *The Physician and the City in Early Modern Europe*, edited by J. Andrew Mendelsohn, Annemarie Kinzelbach, and Ruth Schilling. London: Routledge, 2018.

Leong, Elaine, and Sara Pennell. "Recipe Collections and the Currency of Medical Knowledge in the Early Modern 'Medical Marketplace.'" In *Medicine and the Market in England and Its Colonies, c. 1450–c. 1850*, edited by Mark S. R. Jenner and Patrick Wallis, 133–52. Basingstoke, UK: Palgrave Macmillan, 2007.

Leong, Elaine, and Alisha Rankin. *Secrets and Knowledge in Medicine and Science, 1500–1800*. Farnham, UK: Ashgate, 2011.

———. "Testing Drugs and Trying Cures: Experiment and Medicine in Medieval and Early Modern Europe." *Bulletin of the History of Medicine*, special issue, 91 (2017): 157–82.

Lerman, Nina E., Ruth Oldenziel, and Arwen Mohun, eds. *Gender and Technology: A Reader*. Baltimore: Johns Hopkins University Press, 2003.

Long, Pamela O. *Artisan/Practitioners and the Rise of the New Sciences, 1400–1600*. Corvallis: Oregon State University Press, 2011.

Love, Harold. *The Culture and Commerce of Texts: Scribal Publication in Seventeenth-Century England*. Amherst: University of Massachusetts Press, 1998.

MacDonald, Michael. *Mystical Bedlam: Madness, Anxiety and Healing in Seventeenth-Century England*. Cambridge: Cambridge University Press, 1981.

Malcolm, Noel. "Thomas Harrison and His 'Ark of Studies': An Episode in the History of the Organization of Knowledge." *Seventeenth Century* 19, no. 2 (2004): 196–232.

Marland, Hilary. "'The Doctor's Shop': The Rise of the Chemist and Druggist in Nineteenth-Century Manufacturing Districts." In *From Physick to Pharmacology: Five Hundred Years of British Drug Retailing*, edited by Louise Hill Curth, 79–104. Aldershot, UK: Ashgate, 2006.

Marotti, Arthur. *Manuscript, Print, and the English Renaissance Lyric*. Ithaca, NY: Cornell University Press, 1995.

McConnell, Anita. "Brereton, William, third Baron Brereton of Leighlin (*bap.* 1631, *d.* 1680)." In *Oxford Dictionary of National Biography*. Oxford: Oxford University Press, 2004–. Accessed 19 October 2017. http://www.oxforddnb.com /view/article/39679.

McDermott, James. "Howard, Charles, second Baron Howard of Effingham and first earl of Nottingham (1536–1624), naval commander." In *Oxford Dictionary of National Biography*. Oxford: Oxford University Press, 2004–. Accessed 4 February 2018. https://doi.org/10.1093/ref:odnb/13885.

McKenzie, Donald Francis. *Bibliography and the Sociology of Texts*. London: British Library, 1986.

———. *Making Meaning: "Printers of the Mind" and Other Essays*, edited by Peter D. McDonald and Michael Felix Suarez. Amherst: University of Massachusetts Press, 2002.

McKitterick, David. "'Ovid with a Littleton': The Cost of English Books in the Early Seventeenth Century." *Transactions of the Cambridge Bibliographical Society* 11, no. 2 (1997): 184–234.

McShane, Angela. "Material Culture and 'Political Drinking' in Seventeenth-Century England." *Past and Present* 222, suppl. 9 (2014): 247–76.

McVaugh, Michael. "Determining a Drug's Properties: Medieval Experimental Protocols." *Bulletin of the History of Medicine* 91, no. 2 (2017): 183–209.

———. "The 'Experience-Based Medicine' of the Thirteenth Century." *Early Science and Medicine* 14, no. 1 (2009): 105–30.

———. "The Experimenta of Arnald of Villanova." *Journal of Medieval and Renaissance Studies* 1, no. 1 (1971): 107–18.

———. "Two Montpellier Recipe Collections." *Manuscripta* 20, no. 3 (1976): 175–80.

Meaney, Audrey L. "Variant Versions of Old English Medical Remedies and the Compilation of Bald's *Leechbook*." *Anglo-Saxon England* 13 (1984): 235–68.

Mendelsohn, J. Andrew, and Annemarie Kinzelbach. "Common Knowledge: Bodies, Evidence, and Expertise in Early Modern Germany." *Isis* 108, no. 2 (2017): 259–79.

Mendelson, Sara Heller, and Patricia Crawford. *Women in Early Modern England: 1550–1720.* Oxford: Oxford University Press, 1998.

Milam, Erika Lorraine, and Robert A. Nye, eds. *Scientific Masculinities. Osiris,* special issue, 30 (2015).

Milstein, Joanna. *The Gondi: Family Strategy and Survival in Early Modern France.* Aldershot, UK: Ashgate, 2014.

Mortimer, Ian. *The Dying and the Doctors: The Medical Revolution in Seventeenth-Century England.* Woodbridge: Royal Historical Society/Boydell Press, 2009.

Moss, Ann. *Printed Commonplace-Books and the Structuring of Renaissance Thought.* Oxford: Clarendon Press, 1996.

Mui, Hoh-cheung, and Lorna H. Mui. *Shops and Shopkeeping in Eighteenth-Century England.* Kingston, ON: McGill-Queen's University Press, 1989.

Mukherjee, Ayesha. "The Secrets of Sir Hugh Platt." In *Secrets and Knowledge in Medicine and Science, 1500–1800,* edited by Elaine Leong and Alisha Rankin, 69–86. Aldershot, UK: Ashgate, 2011.

Muldrew, Craig. *The Economy of Obligation: The Culture of Credit and Social Relations in Early Modern England.* Basingstoke, UK: Palgrave Macmillan, 1998.

Munk, William. *The Roll of the Royal College of Physicians of London: Comprising Biographical Sketches.* Vol. 1, *1518–1700*; vol. 2, *1701–1800*; vol. 3, *1801–1825.* London: Royal College of Physicians, 1878.

Munroe, Jennifer. *Gender and the Garden in Early Modern English Literature.* Aldershot, UK: Ashgate, 2008.

———. "Mary Somerset and Colonial Botany: Reading between the Ecofeminist Lines." *Early Modern Studies Journal* 6 (2014): 100–128.

———. "'My Innocent Diversion of Gardening': Mary Somerset's Plants." *Renaissance Studies* 25, no. 1 (2011): 111–23.

Nagy, Doreen Evenden. *Popular Medicine in Seventeenth-Century England.* Bowling Green, OH: Bowling Green State University Popular Press, 1988.

Nance, Brian. *Turquet de Mayerne as Baroque Physician: The Art of Medical Portraiture.* Amsterdam: Rodopi, 2001.

Nappi, Carla. "Bolatu's Pharmacy Trade in Early Modern China." *Early Science and Medicine* 14, no. 6 (2009): 737–64.

Nelson, Paul David. "Lovelace, Francis (c. 1621–1675)." In *Oxford Dictionary of National Biography.* Oxford: Oxford University Press, 2004–. Accessed 19 October 2017. http://www.oxforddnb.com/view/article/17053.

Newcomb, Lori Humphrey. *Reading Popular Romance in Early Modern England.* New York: Columbia University Press, 2002.

Newman, William R., and Lawrence M. Principe. "The Chymical Laboratory Notebooks of George Starkey." In *Reworking the Bench: Research Notebooks in the History of Science*, edited by Frederic L. Holmes, Jürgen Renn, and Hans-Jörg Rheinberger, 25–41. Dordrecht: Kluwer, 2003.

Newton, Hannah. *The Sick Child in Early Modern England, 1580–1720.* Oxford: Oxford University Press, 2012.

North, Marcy. *The Anonymous Renaissance: Cultures of Discretion in Tudor-Stuart England.* Chicago: University of Chicago Press, 2003.

Nummedal, Tara. "Anna Zieglerin's Alchemical Revelations." In *Secrets and Knowledge in Medicine and Science, 1500–1800*, edited by Elaine Leong and Alisha Rankin, 125–42. Farnham, UK: Ashgate, 2011.

Nunn, Hillary. "'Goeing a Broad to Gather and Worke the Flowers': The Domestic Geography of Elizabeth Isham's *My Booke of Remembrance*." In *Ecofeminist Approaches to Early Modernity*, edited by Rebecca Laroche and Jennifer Munroe, 153–74. New York: Palgrave Macmillan, 2011.

Ochs, Kathleen H. "The Royal Society of London's History of Trades Programme: An Early Episode in Applied Science." *Notes and Records of the Royal Society of London* 39, no. 2 (1985): 129–58.

Ogden, Margaret Sinclair. *The 'Liber de Diversis Medicinis' in the Thornton Manuscript (MS Lincoln Cathredral A5.2).* London: Humphrey Milford, 1938.

Ogilvie, Brian W. *The Science of Describing: Natural History in Renaissance Europe.* Chicago: University of Chicago Press, 2006.

O'Hara-May, J. "Foods or Medicines? A Study in the Relationship between Foodstuffs and Materia Medica from the Sixteenth to the Nineteenth Century." *Transactions of the British Society for the History of Pharmacy* 1, no. 2 (1971): 61–97.

Opitz, Donald L., Staffan Bergwik, and Brigitte Van Tiggelen. *Domesticity in the Making of Modern Science.* New York: Springer, 2016.

Osborn, Sally Ann. "The Role of Domestic Knowledge in an Era of Professionalisation: Eighteenth-Century Manuscript Medical Recipe Collections." PhD diss., University of London, 2015.

Östling, Johan, et al. "The History of Knowledge and the Circulation of Knowledge: An Introduction." In *Circulation of Knolwedge: Explorations in the History of Knowledge,* edited by Johan Östling et al., 9–33. Lund, Sweden: Nordic Academic Press, 2018.

Park, Katharine. "Natural Particulars: Medical Epistemology, Practice, and the Literature of Healing Springs." In *Natural Particulars: Nature and the Disciplines in Renaissance Europe*, edited by Anthony Grafton and Nancy G. Siraisi, 347–68. Cambridge, MA: MIT Press, 1999.

———. *Secrets of Women: Gender, Generation, and the Origins of Human Dissection*. New York: Zone Books, 2006.

Pelling, Margaret. "Appearance and Reality: Barber-Surgeons, the Body and Disease." In *London 1500–1700: The Making of the Metropolis*, edited by A. L. Beier and Roger Finlay, 82–112. London: Longman, 1986.

———. "Food, Status and Knowledge: Attitudes to Diet in Early Modern England." In *The Common Lot: Sickness, Medical Occupations and the Urban Poor in Early Modern England*, 39–62. London: Longman, 1998.

———. *Medical Conflicts in Early Modern London: Patronage, Physicians, and Irregular Practitioners, 1550–1640*. Oxford: Clarendon Press, 2003.

———. "Occupational Diversity: Barbersurgeons and the Trades of Norwich, 1550–1640." *Bulletin of the History of Medicine* 56, no. 4 (1982): 484–511.

———. "Thoroughly Resented? Older Women and the Medical Role in Early Modern England." In *Women, Science and Medicine, 1500–1700: Mothers and Sisters of the Royal Society*, edited by Lynette Hunter and Sarah Hutton, 63–88. Thrupp, Stroud, Gloucestershire, UK: Sutton, 1997.

———. "Trade or Profession? Medical Practice in Early Modern England." In *The Common Lot: Sickness, Medical Occupations and the Urban Poor in Early Modern England*, edited by Margaret Pelling, 230–58. London: Longman, 1998.

Pender, Stephen. "Examples and Experience: On the Uncertainty of Medicine." *British Journal for the History of Science* 39, no. 140, pt. 1 (2006): 1–28.

Penman, Leigh T. I. "Omnium Exposita Rapinae: The Afterlives of the Papers of Samuel Hartlib." *Book History* 19, no. 1 (2017): 1–65.

Pennell, Sara. *The Birth of the English Kitchen, 1600–1850*. London: Bloomsbury, 2016.

———. "Making Livings, Lives and Archives: Tales of Four Eighteenth-Century Recipe Books." In *Reading and Writing Recipe Books*, edited by Michelle DiMeo and Sara Pennell, 225–46. Manchester: Manchester University Press, 2013.

———. "Perfecting Practice? Women, Manuscript Recipes and Knowledge in Early Modern England." In *Early Modern Women's Manuscript Writing: Selected Papers from the Trinity/Trent Colloquium*, edited by Victoria Burke and Jonathan Gibson, 237–58. Aldershot, UK: Ashgate, 2004.

Philipot, John, William Ryley, William Harvey, William Harry Rylands, and College of Arms (Great Britain), eds. *The Visitation of the County of Buckingham Made in 1634 by John Philipot, Esq.* Visitation Series 58. London: Harleian Society, 1909.

Pickstone, John V. *Ways of Knowing: A New History of Science, Technology and Medicine*. Manchester: Manchester University Press, 2000.

Pollock, Linda. *With Faith and Physic: The Life of a Tudor Gentlewoman; Lady Grace Mildmay, 1552–1620*. New York: St. Martin's Press, 1995.

Pomata, Gianna. "Observation Rising: Birth of an Epistemic Genre, ca. 1500–1650."

In *Histories of Scientific Observation*, edited by Lorraine Daston and Elizabeth Lunbeck. 55–88. Chicago: University of Chicago Press, 2011.

———. "Praxis Historialis: The Uses of Historia in Early Modern Medicine." In *Historia: Empiricism and Erudition in Early Modern Europe*, edited by Gianna Pomata and Nancy G. Siraisi, 125–37. Cambridge, MA: MIT Press, 2005.

———. "The Recipe and the Case: Epistemic Genres and the Dynamics of Cognitive Practices." In *Wissenschaftsgeschichte und Geschichte des Wissens im Dialog — Connecting Science and Knowledge*, edited by Kaspar von Greyerz, Silvia Flubacher, and Philipp Senn, 131–54. Göttingen: V&R Unipress, 2013.

———. "A Word of the Empirics: The Ancient Concept of Observation and Its Recovery in Early Modern Medicine." *Annals of Science* 68 (2011): 1–25.

Pomata, Gianna, and Nancy G. Siraisi, eds. *Historia: Empiricism and Erudition in Early Modern Europe*. Cambridge, MA: MIT Press, 2005.

Porter, Roy. *Health for Sale: Quackery in England, 1660–1850*. Manchester: University of Manchester Press, 1989.

Postles, David. "The Politics of Address in Early-Modern England." *Journal of Historical Sociology* 18, no. 1–2 (2005): 99–121.

Poynter, F. N. L. "Nicholas Culpeper and His Books." *Journal of the History of Medicine* 17 (1962): 152–67.

Prévost, Michel, and Roman d'Amat. *Dictionnaire de biographie française*. Vol. 8. Paris: Letouzey et Ané, 1959.

Principe, Lawrence. "Sir Kenelm Digby and His Alchemical Circle in 1650s Paris: Newly Discovered Manuscripts." *Ambix* 60 (2013): 3–24.

Pugliano, Valentina. "Pharmacy, Testing and the Language of Truth in Renaissance Italy." *Bulletin of the History of Medicine* 91, no. 2 (2017): 233–73.

Ragland, Evan R. "'Making Trials' in Sixteenth- and Early Seventeenth-Century European Academic Medicine." *Isis* 108 (2017): 503–28.

Rampling, Jennifer M. "Transmuting Sericon: Alchemy as 'Practical Exegesis' in Early Modern England." *Osiris* 29 (2014): 19–34.

Rankin, Alisha. "The Housewife's Apothecary in Early Modern Austria: Wolfgang Helmhard von Hohberg's *Georgica Curiosa* (1682)." *Medicina e Storia* 8, no. 15 (2008): 55–76.

———. "On Anecdote and Antidotes: Poison Trials in Sixteenth-Century Europe." *Bulletin of the History of Medicine* 91, no. 2 (2017): 274–302.

———. *Panaceia's Daughters: Noblewomen as Healers in Early Modern Germany*. Chicago: University of Chicago Press, 2013.

Raphael, Renée. *Reading Galileo: Scribal Technologies and the Two New Sciences*. Baltimore: Johns Hopkins University Press, 2017.

Raven, James, Helen Small, and Naomi Tadmor, eds. *The Practice and Representation of Reading in England*. Cambridge: Cambridge University Press, 1996.

Ray, Meredith K. *Daughters of Alchemy: Women and Scientific Culture in Early Modern Italy*. Cambridge, MA: Harvard University Press, 2015.

Raymond, Joad. *Pamphlets and Pamphleteering in Early Modern Britain*. Cambridge: Cambridge University Press, 2006.

Rees, Graham. "An Unpublished Manuscript by Francis Bacon: *Sylva Sylvarum* Drafts and Other Working Notes." *Annals of Science* 38, no. 4 (1981): 377–412.

Reid, John G. "Temple, Sir Thomas, first Baronet (*bap.* 1614, *d.* 1674)." In *Oxford Dictionary of National Biography*. Oxford: Oxford University Press, 2004–. Accessed 19 October 2017. http://www.oxforddnb.com/view/article/27120.

Reinke-Williams, Tim. *Women, Work and Sociability in Early Modern London*. Basingstoke, UK: Palgrave Macmillan, 2014.

Richards, Jennifer, and Fred Schurink. "The Textuality and Materiality of Reading in Early Modern England." *Huntington Library Quarterly*, special issue, 73, no. 3 (2010): 345–61.

Richardson, R. C. *Household Servants in Early Modern England*. Manchester: Manchester University Press, 2010.

Robbins, Rossell Hope. "Medical Manuscripts in Middle English." *Speculum* 45, no. 3 (1970): 393–415.

Roberts, Lissa L., Simon Schaffer, and Peter Dear, eds. *The Mindful Hand: Inquiry and Invention from the Late Renaissance to Early Industrialisation*. Amsterdam: Edita Koninklijke Nederlandse Akademie van Wetenschappen, 2007.

Rosenberg, Daniel, ed. "Early Modern Information Overload." *Journal of the History of Ideas* 64 (2003).

Rusu, Doina-Cristina, and Christoph Lüthy. "Extracts from a Paper Laboratory: The Nature of Francis Bacon's *Sylva Sylvarum*." *Intellectual History Review* 27, no. 2 (2017): 171–202.

Sambrook, Pamela. *Country House Brewing in England, 1500–1900*. London: Hambledon Press, 1996.

Sawyer, Ronald C. "Patients, Healers, and Disease in the Southeast Midlands, 1597–1634." PhD diss., University of Wisconsin–Madison, 1986.

Schaffer, Simon. "Glass Works: Newton's Prisms and the Uses of Experiment." In *The Uses of Experiment: Studies in the Natural Sciences*, edited by David Gooding, Trevor Pinch, and Simon Schaffer, 67–104. Cambridge: Cambridge University Press, 1989.

Schickore, Jutta. "The Significance of Re-doing Experiments: A Contribution to Historically Informed Methodology." *Erkenntnis* 75, no. 3 (2011): 325–47.

———. "Trying Again and Again: Multiple Repetitions in Early Modern Reports of Experiments on Snake Bites." *Early Science and Medicine* 15 (2010): 567–617.

Schiebinger, Londa. *The Mind Has No Sex? Women in the Origins of Modern Science*. Cambridge: Cambridge University Press, 1989.

Schmitt, Charles B. "Experience and Experiment: A Comparison of Zabarella's View with Galileo's in *De Motu*." *Studies in the Renaissance* 16 (1969): 80–138.

Scott-Warren, Jason. "Early Modern Bookkeeping and Life-Writing Revisited: Accounting for Richard Stonley." *Past and Present* 230, suppl. 11 (2016): 151–70.

———. *Sir John Harington and the Book as Gift*. Oxford: Oxford University Press, 2001.

Seaward, Paul. "Dering, Sir Edward, second baronet (1625–1684), politician." In *Oxford Dictionary of National Biography*. Oxford: Oxford University Press, 2004–. Accessed 4 February 2018. https://doi.org/10.1093/ref:odnb/37354.

Shapin, Steven. "The House of Experiment in Seventeenth-Century England." *Isis* 79, no. 3 (1988): 373–404.

———. "The Invisible Technician." *American Scientist* 77, no. 6 (1989): 554–63.

———. *The Social History of Truth: Civility and Science in Seventeenth-Century England*. Chicago: University of Chicago Press, 1994.

Shapin, Steven, and Simon Schaffer. *Leviathan and the Air-Pump: Hobbes, Boyle, and the Experimental Life*. Princeton, NJ: Princeton University Press, 1985.

Sharpe, Kevin M. *Reading Revolutions: The Politics of Reading in Early Modern England*. New Haven, CT: Yale University Press, 2000.

Sharpe, Kevin, and Steven N. Zwicker, eds. *Reading, Society and Politics in Early Modern England*. Cambridge: Cambridge University Press, 2003.

Sherman, William Howard. *John Dee: The Politics of Reading and Writing in the English Renaissance*. Amherst: University of Massachusetts Press, 1995.

———. *Used Books: Marking Readers in Renaissance England*. Philadelphia: University of Pennsylvania Press, 2008.

———. "What Did Renaissance Readers Write in Their Books?" In *Books and Readers in Early Modern England: Material Studies*, edited by Jennifer Andersen and Elizabeth Sauer, 119–37. Philadelphia: University of Pennsylvania Press, 2002.

Shirley, J. W. "The Scientific Experiments of Sir Walter Raleigh, the Wizard Earl and the Three Magi in the Tower, 1603–1617." *Ambix* 4 (1949–51): 52–66.

Sibum, Heinz Otto. "Reworking the Mechanical Value of Heat: Instruments of Precision and Gestures of Accuracy in Early Victorian England." *Studies in History and Philosophy of Science, Part A,* 26, no. 1 (1995): 73–106.

Siena, Kevin P. "The 'Foul Disease' and Privacy: The Effects of Venereal Disease and Patient Demand on the Medical Marketplace in Early Modern London." *Bulletin of the History of Medicine* 75, no. 2 (2001): 199–224.

Slack, Paul. *The Impact of Plague in Tudor and Stuart England*. Oxford: Oxford University Press, 1985.

———. "Mirrors of Health and Treasures of Poor Men: The Uses of the Vernacular Medical Literature of Tudor England." In *Health, Medicine and Mortality in the Sixteenth Century*, edited by Margaret Pelling and Charles Webster, 237–73. Cambridge: Cambridge University Press, 1979.

Smallwood, Frank T. "Lady Johanna St. John as a Medical Practitioner." *Friends of Lydiard Tregoz Report* 6 (1973): 13–33.

———. "The Will of Dame Johanna St. John." *Notes and Queries* 16, no. 9 (1969): 344–47.

Smith, Daniel Starza. "'La conquest du sang real': Edward, Second Viscount Conway's Quest for Books." In *From Compositor to Collector: Essays on the Book Trade*, edited by Matthew Day and John Hinks, 191–216. London: British Library/Print Networks, 2012.

———. *John Donne and the Conway Papers: Patronage and Manuscript Circulation in the Early Seventeenth Century*. Oxford: Oxford University Press, 2014.

Smith, Lisa. "The Relative Duties of a Man: Domestic Medicine in England and France, ca. 1685–1740." *Journal of Family History* 31, no. 3 (2006): 237–56.

———. "Sir Hans Sloane, Friend and Physician to the Family." In *From Books to Bezoars: Sir Hans Sloane, Collector of an Age*, edited by Alison Walker, Michael Hunter, and Arthur MacGregor, 48–56. London: British Library Publications, 2012.

Smith, Pamela. *The Body of the Artisan: Art and Experience in the Scientific Revolution*. Chicago: University of Chicago Press, 2004.

———. "In the Workshop of History: Making, Writing, and Meaning." *West 86th: A Journal of Decorative Arts, Design History, and Material Culture, Shaping Objects; Art, Materials, Making, and Meanings in the Early Modern World* 19 (2012): 4–31.

———. "Science on the Move: Recent Trends in the History of Early Modern Science." *Renaissance Quarterly* 62 (2009): 345–75.

———. "What Is a Secret? Secrets and Craft Knowledge in Early Modern Europe." In *Secrets and Knowledge in Medicine and Science, 1500–1800*, edited by Elaine Leong and Alisha Rankin, 47–66. Aldershot, UK: Ashgate, 2011.

Smith, Pamela H., Amy R. W. Meyers, and Harold J. Cook, eds. *Ways of Making and Knowing: The Material Culture of Empirical Knowledge*. Bard Graduate Center Cultural Histories of the Material World. Ann Arbor: University of Michigan Press, 2014.

Smith, Pamela H., and Benjamin Schmidt, eds. *Making Knowledge in Early Modern Europe: Practices, Objects, and Texts, 1400–1800*. Chicago: University of Chicago Press, 2007.

Smyth, Adam. *Autobiography in Early Modern England*. Cambridge: Cambridge University Press, 2010.

Snook, Edith. "'Soveraigne Receipts' and the Politics of Beauty in *The Queens Closet Opened*." *Early Modern Literary Studies*, special issue, 15 (2007): 1–19.

Soll, Jacob. *The Information Master: Jean-Baptiste Colbert's Secret State Intelligence System: Cultures of Knowledge in the Early Modern World*. Ann Arbor: University of Michigan Press, 2009.

Spalding, Ruth. "Winston, Thomas (c. 1575–1655)." In *Oxford Dictionary of Na-*

tional Biography. Oxford: Oxford University Press, 2004–. Accessed 19 October 2017. http://www.oxforddnb.com/view/article/29762.

Spiller, Elizabeth. "Introduction." In *Seventeenth-Century English Recipe Books: Cooking, Physic and Chirurgery in the Works of Elizabeth Talbot Grey and Aletheia Talbot Howard. Essential Works for the Study of Early Modern Women, ser. 3, pt. 3, vol. 3*, edited by Elizabeth Spiller, ix–li. Aldershot, UK: Ashgate, 2008.

———. "Printed Recipe Books in Medical, Political, and Scientific Contexts." In *The Oxford Handbook of Literature and the English Revolution*, edited by Laura Knoppers, 516–33. Oxford: Oxford University Press, 2012.

———. "Recipes for Knowledge: Maker's Knowledge Traditions, Paracelsian Recipes, and the Invention of the Cookbook, 1600–1660." In *Renaissance Food from Rabelais to Shakespeare: Cultural Readings and Cultural Histories*, edited by Joan Fitzpatrick, 55–72. Aldershot, UK: Ashgate, 2010.

Spufford, Margaret. "First Steps in Literacy: The Reading and Writing Experiences of the Humblest Seventeenth-Century Autobiographers." *Social History* 4 (1979): 407–35.

———. *Small Books and Pleasant Histories: Popular Fiction and Its Readership in Seventeenth-Century England*. Cambridge: Cambridge University Press, 1981.

Squibb, G. D. [George Drewry], Edward Bysshe, and the Harleian Society, eds. *The Visitation of Somerset and the City of Bristol 1672: Made by Sir Edward Bysshe, Knight, Clarenceux King of Arms/Transcribed and Ed. G. D. Squibb*. London: Harleian Society, 1992.

———. *The Visitation of Oxfordshire 1669 and 1675: Made by Sir Edward Bysshe, Knight, Clarenceux King of Arms/Transcribed and Edited by G. D. Squibb*. London: Harleian Society, 1993.

Stannard, J. "Rezeptliteratur as Fachliteratur." In *Studies on Medieval Fachliteratur*, edited by W. Eamon, 59–73. Brussels: Omirel, 1982.

Starkey, George. *Alchemical Laboratory Notebooks and Correspondence*. Edited by William R. Newman and Lawrence M. Principe. Chicago: University of Chicago Press, 2004.

Stevenson, J., and P. Davidson, "Introduction." In *The Closet of the Eminently Learned Sir Kenelm Digby Kt. Opened*, edited by Jane Stevenson and Peter Davidson. Totnes, Devon, UK: Prospect, 1997.

Stine, Jennifer K. "Opening Closets: The Discovery of Household Medicine in Early Modern England." PhD diss., Stanford University, 1996.

Stobart, Anne. *Household Medicine in Seventeenth-Century England*. London: Bloomsbury, 2016.

———. "The Making of Domestic Medicine: Gender, Self-Help and Therapeutic Determination in Household Healthcare in South-West England in the Late Seventeenth Century." PhD diss., Middlesex University, 2008.

Stobart, Jon. *Sugar and Spice: Grocers and Groceries in Provincial England, 1650–1830*. Oxford: Oxford University Press, 2013.

Stolberg, Michael. "Learning from the Common Folks: Academic Physicians and Medical Lay Culture in the Sixteenth Century." *Social History of Medicine* 27, no. 4 (2014): 649–67.

———. "Medizinische Loci communes: Formen und Funktionen einer ärztlichen Aufzeichnungspraxis im 16. und 17. Jahrhundert." *NTM Zeitschrift für Geschichte der Wissenshaften, Teknik und Medizin* 21 (2013): 37–60.

———. "'You Have No Good Blood in Your Body': Oral Communication in Sixteenth-Century Physicians' Medical Practice." *Medical History* 59, no. 1 (2015): 63–82.

Strocchia, Sharon, ed. "Women and Healthcare in Early Modern Europe." *Renaissance Studies*, special issue, 28, no. 4 (2014): 579–96.

Sumner, James. *Brewing Science, Technology and Print, 1700–1880*. London: Routledge, 2013.

Sweetman, George. *The History of Wincanton, Somerset, from the Earliest Times to the Year 1903*. London: Henry Williams, 1903.

Tadmor, Naomi. *Family and Friends in Eighteenth-Century England: Household, Kinship, and Patronage*. Cambridge: Cambridge University Press, 2001.

Terrall, Mary. *Catching Nature in the Act: Réaumur and the Practice of Natural History in the Eighteenth Century*. Chicago: University of Chicago Press, 2013.

———. "Masculine Knowledge, the Public Good, and the Scientific Household of Réaumur." *Osiris* 30 (2015): 182–201.

Theophano, Janet. *Eat My Words: Reading Women's Lives through the Cookbooks They Wrote*. New York: Palgrave, 2002.

Thirsk, Joan. *Food in Early Modern England: Phases, Fads, Fashions, 1500–1760*. London: Bloomsbury Academic, 2007.

Totelin, Laurence. *Hippocratic Recipes: Oral and Written Transmission of Pharmacological Knowledge in Fifth- and Fourth-Century Greece*. Leiden: Brill, 2009.

Touwaide, Alain. "Quid pro Quo: Revisiting the Practice of Substitution in Ancient Pharmacy." In *Herbs and Healing, from the Ancient Mediterranean through the Medieval West*, edited by Ann Van Arsdall and Timothy Graham, 19–61. Farnham, UK: Ashgate, 2012.

Trevor-Roper, Hugh. "Mayerne, Sir Theodore Turquet de (1573–1655)." In *Oxford Dictionary of National Biography*. Oxford: Oxford University Press, 2004. Accessed 19 October 2017. http://www.oxforddnb.com/view/article/18430.

Turnbull, George Henry. *Hartlib, Dury and Comenius: Gleanings from Hartlib's Papers*. Liverpool: University Press of Liverpool, 1947.

———. *Samuel Hartlib: A Sketch of His Life and His Relations to J. A. Comenius*. Oxford: Oxford University Press, 1920.

Valleriani, Matteo, ed. *The Structures of Practical Knowledge*. Heidelberg: Springer, 2017.

Venn, John, and John Archibald Venn. *Alumni Cantabrigienses: A Biographical List of All Known Students, Graduates and Holders of Office at the University of Cambridge, from the Earliest Times to 1900.* 10 vols. Cambridge: Cambridge University Press, 1922–54.

Vine, Angus. "Commercial Commonplacing: Francis Bacon, the Waste-Book, and the Ledger." In *Manuscript Miscellanies, 1450–1700,* edited by Richard Beadle, Peter Beal, Colin Burrow, and A. S. G. Edwards, 197–218. London: British Library, 2011.

von Oertzen, Christine, Maria Rentezi, and Elizabeth Watkins, eds. "Beyond the Academy: Histories of Gender and Knowledge." *Centaurus* 55, no. 2 (2013).

Wall, Wendy. "Literacy and the Domestic Arts." *Huntington Library Quarterly* 73, no. 3 (2010): 383–412.

———. *Recipes for Thought: Knowledge and Taste in the Early Modern English Kitchen.* Material Texts. Philadelphia: University of Pennsylvania Press, 2016.

Wallis, Patrick, ed. "Changes in Medical Care." *Journal of Social History*, special issue, 49, no. 2 (2016).

———. "Consumption, Retailing, and Medicine in Early-Modern London." *Economic History Review* 16 (2008): 26–53.

Walsham, Alexandra. "The Social History of the Archive: Record-Keeping in Early Modern Europe." *Past and Present* 230, suppl. 11 (1 January 2016): 9–48.

Ward, Jenny. "Brugis, Thomas (*b.* in or before 1620, *d.* in or after 1651)." In *Oxford Dictionary of National Biography.* Oxford: Oxford University Press, 2004–. Accessed 19 October 2017. http://www.oxforddnb.com/view/article/3769.

Wear, Andrew. *Knowledge and Practice in English Medicine, 1550–1680.* Cambridge: Cambridge University Press, 2000.

———. "Popularized Ideas of Health and Illness in Seventeenth-Century France." *Seventeenth-Century French Studies* 8, no. 1 (1986): 229–42.

Webster, Charles. *The Great Instauration: Science, Medicine and Reform, 1626–1660.* London: Duckworth, 1975.

Weisser, Olivia. *Ill Composed: Sickness, Gender and Belief in Early Modern England.* New Haven, CT: Yale University Press, 2015.

Whittle, Jane. "Enterprising Widows and Active Wives: Women's Unpaid Work in the Household Economy of Early Modern England." *History of the Family* 19, no. 3 (2014): 283–300.

———. "The House as a Place of Work in Early Modern Rural England." *Home Cultures* 8, no. 2 (2011): 133–50.

———. "Housewives and Servants in Rural England, 1440–1650: Evidence of Women's Work from Probate Documents." *Transactions of the Royal Historical Society*, 6th ser., 15 (2005): 51–74.

Whittle, Jane, and Elizabeth Griffiths. *Consumption and Gender in the Early Seventeenth-Century Household: The World of Alice Le Strange.* Oxford: Oxford University Press, 2012.

Wilson, Bronwen. "Social Networking: The *Album Amicorum* and Early Modern Public-Making." In *Beyond the Public Sphere: Opinions, Publics, Spaces in*

Early Modern Europe, edited by Massimo Rospocher, 205–23. Bologna: Società editrice il Mulino, 2012.

Wilson, C. Anne. *Food and Drink in Britain: From the Stone Age to Recent Times*. London: Constable, 1973.

Withey, Alun. "Crossing the Boundaries: Domestic Recipe Collections in Early Modern Wales." In *Reading and Writing Recipe Books*, edited by Michelle DiMeo and Sara Pennell, 179–202. Manchester: Manchester University Press, 2013.

———. *Physick and the Family: Health, Medicine and Care in Wales, 1600–1750*. Manchester: Manchester University Press, 2011.

Withington, Phil. "Company and Sociability in Early Modern England." *Social History* 32, no. 3 (2007): 291–307.

Woolf, Daniel. *The Social Circulation of the Past: English Historical Culture, 1500–1730*. Oxford: Oxford University Press, 2003.

Woudhuysen, Henry R. *Sir Philip Sidney and the Circulation of Manuscripts, 1558–1640*. Oxford: Oxford University Press, 1996.

Wright, Stephen. "Fane, Mildmay, second Earl of Westmorland (1602–1666)." In *Oxford Dictionary of National Biography*. Oxford: Oxford University Press, 2004–. Accessed 19 October 2017. http://www.oxforddnb.com/view/article/9139.

———. "Wright, Abraham (1611–1690)." In *Oxford Dictionary of National Biography*. Oxford: Oxford University Press, 2004–. Accessed 19 October 2017. http://www.oxforddnb.com/view/article/30027.

Wrightson, Keith. "The 'Decline of Neighbourliness' Revisited." In *Local Identities in Late Medieval and Early Modern England*, edited by Norman L. Jones and Daniel Woolf, 19–49. Basingstoke, UK: Palgrave Macmillan, 2007.

Yale, Elizabeth. "The History of Archives: The State of the Discipline." *Book History* 18, no. 1 (2015): 332–59.

———. "Marginalia, Commonplaces, and Correspondence: Scribal Exchange in Early Modern Science." *Studies in History and Philosophy of Science, Part C: Studies in History and Philosophy of Biological and Biomedical Sciences* 42, no. 2 (2011): 193–202.

———. *Sociable Knowledge: Natural History and the Nation in Early Modern Britain*. Philadelphia: University of Pennsylvania Press, 2016.

Yeo, Richard. *Notebooks, English Virtuosi, and Early Modern Science*. Chicago: University of Chicago Press, 2014.

———, ed. "Note-Taking in Early Modern Europe." *Intellectual History Review* 20, no. 3 (2010).

Index